SpringerBriefs in Mathematics

SpringerBriefs in Mathematics showcases expositions in all areas of mathematics and applied mathematics. Manuscripts presenting new results or a single new result in a classical field, new field, or an emerging topic, applications, or bridges between new results and already published works, are encouraged. The series is intended for mathematicians and applied mathematicians.

More information about this series at http://www.springer.com/series/10030

Antonio Cañada • Salvador Villegas

A Variational Approach to Lyapunov Type Inequalities

From ODEs to PDEs

 Springer

Antonio Cañada
Department of Mathematical Analysis
University of Granada
Granada, Spain

Salvador Villegas
Department of Mathematical Analysis
University of Granada
Granada, Spain

ISSN 2191-8198　　　　ISSN 2191-8201　(electronic)
SpringerBriefs in Mathematics
ISBN 978-3-319-25287-2　　ISBN 978-3-319-25289-6　(eBook)
DOI 10.1007/978-3-319-25289-6

Library of Congress Control Number: 2015953446

Mathematics Subject Classification (2010): 34B05, 34B15, 34C10, 34L15, 35J20, 35J25, 35J65, 49R05

Springer Cham Heidelberg New York Dordrecht London

Printed on acid-free paper

Springer International Publishing AG Switzerland is part of Springer Science+Business Media (www.springer.com)

To Ery and Patricia

Foreword

In his 1892 Doctoral thesis at Kharkov University on *The general problem of the stability of motion*, Aleksandr M. Lyapunov considered the stability of Hill's equation

$$x'' + a(t)x = 0, \tag{1}$$

where $a(t)$ is continuous and L-periodic. If we write $a \prec b$ when $a(t) \leq b(t)$ for all $t \in [0, L]$ and $a(t) < b(t)$ on a subset of positive measure, Lyapunov proved that if $0 \prec a$ and

$$\int_0^L a \leq \frac{4}{L}, \tag{2}$$

then all solutions of (1) are such that

$$x(L) = e^{\pm i\theta} x(0), \quad x'(L) = e^{\pm i\theta} x'(0) \tag{3}$$

for some $\theta > 0$ such that $\sin\theta \neq 0$ and, in particular, are bounded over **R**. This is the first occurrence of what is called today *Lyapunov inequality*.

As variational equations for periodic solutions of nonlinear oscillators are of type (1), Lyapunov's result was important in the study of their stability. Many improvements, refinements, and generalizations of condition (2) followed almost immediately in Russia and elsewhere. An excellent description and bibliography can be found in Cesari's monograph *Asymptotic Behavior and Stability Problems in Ordinary Differential Equations*, Springer, 1963, and in Yakubovitch–Starzhinskii's two volumes of *Linear Differential Equations with Periodic Coefficients*, Halsted Press, 1975.

But inequality (2) is useful in other contexts than stability. Dealing in 1956 with the second-order nonlinear equation

$$x'' + R(t, x, x')x = Q(t, x, x') \tag{4}$$

with R, Q L-periodic with respect to t, continuous, and Q is bounded on $\mathbf{R}^3$, Volpato (*Rend. Sem. Mat. Univ. Padova* 25, 371–385) proved the existence of an L-periodic solution for (4) when $0 \leq p(t) \leq R(t, x, y) \leq P(t)$, for all $(t, x, y) \in \mathbf{R}^3$ and some continuous L-periodic functions $0 \leq p(t) \leq P(t)$ such that $0 \prec p$ and

$$\int_0^L P \leq \frac{4}{L}. \tag{5}$$

One recognizes in (5) Lyapunov inequality (2) for P. Those conditions imply that whenever a is continuous and L-periodic and such that $p(t) \leq a(t) \leq P(t)$ for all $t \in \mathbf{R}$, all solutions of (1) satisfy (3), and hence 0 is its only L-periodic solution. In 1964, Lasota and Opial (*Ann. Polon. Math.* 16, 69–94) replaced 4 by 16 in condition (5) and showed that 16 is the best possible constant. The existence of an L-periodic solution of (4) follows from a combination of those linear results with Schauder's fixed point theorem.

The existence of an L-periodic solution for (4) has been proved later when $p(t) \leq R(t, x, y) \leq P(t)$, for all $(t, x, y) \in \mathbf{R}^3$, with p and P such that $0 < \int_0^L p$, and

$$P \prec \frac{4\pi^2}{L^2} \tag{6}$$

or

$$\frac{4\pi^2 n^2}{L^2} \prec p \leq P \prec \frac{4\pi^2(n+1)^2}{L^2} \tag{7}$$

for some integer $n \geq 1$.

Inequality (2) or (5) can be seen as L^1-norm conditions on a or P and inequality (6) or (7) as L^∞-norm conditions on p and P. One can therefore think about the possibility of Lyapunov-type inequalities in L^p-norm $\|\cdot\|_p$ ($1 \leq p \leq \infty$) and of a possible unified treatment. One can also raise similar questions for other boundary conditions, like Dirichlet, Neumann, or anti-periodic ones, and investigate the possibility of extending the results to partial differential equations or systems of ordinary differential equations.

All those questions are elegantly and effectively answered in the monograph *A Variational Approach to Lyapunov-Type Inequalities* of Antonio Cañada and Salvador Villegas. Putting the main emphasis on Neumann boundary conditions

$$x'(0) = x'(L) = 0 \tag{8}$$

without neglecting other ones, they introduce and use an original and powerful variational approach to obtain the best L^p-*Lyapunov constant* for (1)–(8), namely,

$$\beta_p := \inf\{\|a^+\|_p : a \in L^p(0,L) \setminus \{0\}, \int_0^L a \geq 0,$$

(1)–(8) has nontrivial solutions}

Interestingly, β_p is achieved except for $p = 1$, provides the expected values $\beta_\infty = \frac{\pi^2}{L^2}$, $\beta_1 = \frac{4}{L}$, and an explicit formula is given for $p \in (1, \infty)$. Characterizations in terms of Rayleigh-type quotients are obtained as well. Those results, and similar ones for other boundary conditions, imply various existence and uniqueness results for corresponding nonlinear problems.

In order to generalize conditions of type (7) for the L^1-norm, the idea is nicely extended to higher order eigenvalues by introducing, for any integer $n \geq 1$, the L^1-*Lyapunov constants*

$$\beta_{1,n} = \inf\{a \in L^1(0,L) : \lambda_n = \frac{n^2\pi^2}{L^2} \prec a, (1)–(8) \text{ has nontrivial solutions}\}$$

If the determination of the β_p was done using essentially tools of the calculus of variations, that of the (non-achieved) $\beta_{1,n}$ is a striking combination of calculus of variations and hard analysis.

L^1 is rarely a good space for partial differential equations, and the fact is confirmed once more here. When defining and computing Lyapunov constant β_p for the Neumann problem

$$- \Delta u(x) = a(x)u(x) \text{ in } \Omega, \quad \frac{\partial u}{\partial n} = 0 \text{ on } \partial\Omega \qquad (9)$$

with $\Omega \subset \mathbf{R}^N$ ($N \geq 2$) a bounded regular domain and $a \in L^p(\Omega)$, the authors are faced with new situations, depending upon the relation between p and N. Indeed, $\beta_1 = 0$ for each $N \geq 2$, $\beta_p > 0$ for all $p \in (1, \infty]$ when $N = 2$ and, for $N \geq 3$, $\beta_p > 0$ if and only if $p \geq N/2$. Furthermore, β_p is achieved when $p > N/2$. A fine analysis of the dependence of β_p with respect to p is given, and the radial case provides refined results that are unreachable in the general case.

Everybody having experienced the use of variational methods knows that when they work for scalar differential equations, they work in general for some classes of differential systems. The last chapter of the book of Cañada and Villegas nicely illustrates this fact.

Mathematicians interested in the qualitative theory of linear differential equations will find in this monograph a number of interesting applications of Lyapunov inequalities and constants to the important question of disfocality. On the other hand, besides their mathematical interest, for example, in the theory of inverse problems, the linear equations with periodic coefficients have found important applications in quantum mechanics, since the pioneering work of Léon Brillouin.

Consequently, the book of Cañada and Villegas will be of interest for a substantial part of the mathematical community, from analysts to mathematical physicists. They will find there a modern, original, and elegant treatment of problems which, as hopefully shown by this Foreword, have their roots in classical papers on differential equations and are treated, in an elegant style, using the most recent techniques of functional analysis and the calculus of variations.

Louvain-la-Neuve, Belgium   Jean Mawhin
April 2015

Preface

Different problems make the study of the so-called Lyapunov-type inequalities of great interest, both in pure and applied mathematics. Although the original historical motivation was the study of the stability properties of Hill's equation, other questions that arise in systems at resonance, crystallography, isoperimetric problems, Rayleigh-type quotients, oscillation and intervals of disconjugacy, etc. lead to the study of this type of inequalities for differential equations. This classical area of mathematics plays a significant role in the current research and remains a source of inspiration to this day.

In this book we examine in a detailed way some of the main aspects of this topic, including the most relevant results obtained by the authors in the last 12 years, as well as many other related results. Obviously, the selection of material is partly conditioned by the interest of the authors.

In our opinion, the contents of the book concerning higher eigenvalues, partial differential equations, and systems of equations are particularly innovative and through the whole monograph, an especial emphasis is done in the variational characterization of the best Lyapunov constants. This unified variational point of view makes possible the study of many cases, featuring a systematic discussion of different types of equations and boundary conditions, both for ordinary and partial differential equations. The applications include nonlinear resonant problems, the study of the stability of linear periodic equations (both for scalar and systems of equations), and the analysis of the sign of the eigenvalues of certain eigenvalue problems.

This work can be considered self-contained, with detailed proofs and a special emphasis on motivation and understanding of the basic ideas. Taking in mind a balanced presentation of both pure and applied aspects, we have tried to write this work in a style accessible to a broad audience, although a great variety of methods from classical analysis, differential equations, and nonlinear functional analysis are used. However, some proofs (especially those referring to the PDE case) are particularly laborious.

The book is addressed to experienced researchers working in the subject and to young researchers who want to start on these topics. The expository content, with detailed proofs and an appropriate list of references in each chapter, brings the reader quickly to the forefront of research. The volume contains numerous explanatory notes on the showed results and their relation to the existing literature.

Granada, Spain   Antonio Cañada
April 2015   Salvador Villegas

Acknowledgments

Our deepest gratitude to our friend and colleague Jean Mawhin for writing the foreword. His influence on the research group on differential equations at the Department of Mathematical Analysis of the University of Granada has been, along 35 years, decisive in all the aspects. We are also very grateful to our colleagues: G. López, J.A. Montero, and P. Torres who read this text and made useful comments and suggestions. We want to thank the referees for their constructive comments. The editorial and production staff of Springer have always been ready to offer a very efficient assistance. Finally, this work has been supported by the project MTM2012.37960 of the Spanish Science and Innovation Ministry.

Contents

Acronyms and Notation

N	The set of natural numbers
R	The set of real numbers
b.v.p.	Boundary value problem
ODEs	Ordinary differential equations
PDEs	Partial differential equations
$L^p(0,L)$ $1 \leq p < \infty$	The Lebesgue space of measurable functions $a(\cdot)$ such that $|a(\cdot)|^p$ is integrable in $(0,L)$
$L^\infty(0,L)$	The Lebesgue space of measurable functions such that there exists a constant c satisfying $|a(x)| \leq c$, almost everywhere (a.e.) in $(0,L)$
$L^p(\Omega)$ $1 \leq p < \infty$	The Lebesgue space of measurable functions $a(\cdot)$ such that $|a(\cdot)|^p$ is integrable in Ω, a bounded and regular domain in $\mathbf{R}^N$
$L^\infty(\Omega)$	The Lebesgue space of measurable functions such that there exists a constant c satisfying $|a(x)| \leq c$, almost everywhere (a.e.) in Ω
$L_T(\mathbf{R}, \mathbf{R})$	The set of T-periodic functions $a(\cdot)$ such that $a|_{(0,T)} \in L^1(0,T)$
$\\| \cdot \\|_p$	The usual norm in the spaces L^p
$\langle \cdot, \cdot \rangle$	The usual scalar product in $\mathbf{R}^n$
$H^1(0,L)$	The usual Sobolev space on the interval $(0,L)$
$H^1(\Omega), H^1_0(\Omega)$	The usual Sobolev spaces on a bounded and regular domain $\Omega \subset \mathbf{R}^N$
$\dfrac{\partial}{\partial n}$	Outer normal derivative on $\partial\Omega$
$W^{m,p}(\Omega)$	The usual Sobolev space on a bounded and regular domain $\Omega \subset \mathbf{R}^N$
$H^1_T(0,T)$	The subset of T-periodic functions of the Sobolev space $H^1(0,T)$
B_r	The open ball in $\mathbf{R}^N$ of center zero and radius r

$C_T(\mathbf{R}, \mathbf{R})$	The set of real T-periodic and continuous functions defined in $\mathbf{R}$
$\mathscr{M}(\mathbf{R})$	The set of real $n \times n$ matrices
β_p	Best L_p Lyapunov constant
f_u	Partial derivative of the function f with respect to u
$c \prec d$	$c, d \in L^1(0, L)$, $c(x) \leq d(x)$ for a.e. $x \in [0, L]$ and $c(x) < d(x)$ on a set of positive measure
$C \leq D$	The relation $C \leq D$ between $n \times n$ matrices means that $D - C$ is positive semi-definite
a^+	The positive part of the function a, i.e., $a^+(t) = \max\{0, a(t)\}$
$\det A$	The determinant of the matrix A
$\rho(A)$	The spectral radius of the matrix A
Trace (A)	The trace of the matrix A
Δ	The Laplacian operator

Chapter 1
Introduction

Abstract In this chapter we briefly present some historical motivations which make the study of the so-called Lyapunov-type inequalities of great interest, both in pure and applied mathematics. In particular, three topics are highlighted: the stability properties of the Hill's equation, the study of the sign of the eigenvalues of certain eigenvalue problems, and the analysis of nonlinear resonant problems. After, we describe the contents of the book.

1.1 Stability, Resonance, and Lyapunov Inequalities

The Hill's equation

$$u''(t) + a(t)u(t) = 0, \ t \in \mathbf{R},$$ (1.1)

where the function a satisfies

$$a : \mathbf{R} \to \mathbf{R} \ \text{is} \ T - \text{periodic and} \ a \in L^1(0, T)$$ (1.2)

models many phenomena in applied sciences [8, 9, 14].

The study of the stability properties of (1.1) is of special interest. Whenever all solutions of (1.1) are bounded, we say that (1.1) is stable; otherwise we say that it is unstable. Floquet theory assures that such stability properties depend on characteristic multipliers [8, 9].

In the book by W. Magnus and S. Winkler ([14], Chap. V), the authors present in detail the conclusions shown by G. Borg in [2]. Among other things, the following result is mentioned (in what follows, we denote by $L_T(\mathbf{R}, \mathbf{R})$ the set of functions $a(\cdot)$ satisfying (1.2)):

If $a \in L_T(\mathbf{R}, \mathbf{R})$ and for some $n \in \mathbf{N} \bigcup \{0\}$

$$\frac{n^2\pi^2}{T^2} \prec a \prec \frac{(n+1)^2\pi^2}{T^2}$$ (1.3)

then (1.1) is stable.

© The Author(s) 2015
A. Cañada, S. Villegas, *A Variational Approach to Lyapunov Type Inequalities*,
SpringerBriefs in Mathematics, DOI 10.1007/978-3-319-25289-6_1

Here, for $c, d \in L^1(0, T)$, we write $c \prec d$ if $c(t) \leq d(t)$ for a.e. $t \in [0, T]$ and $c(t) < d(t)$ on a set of positive measure.

The hypothesis (1.3) can be seen as an $L_\infty - L_\infty$ restriction, in the sense that the function $a(\cdot)$ is bounded, from above and from below, by appropriate constants.

In the early twentieth century Lyapunov provided an $L_\infty - L_1$ restriction [13, 14]: if

$$0 \prec a, \quad \int_0^T a(t)\, dt \leq \frac{4}{T} \tag{1.4}$$

then (1.1) is stable.

Clearly, (1.3) and (1.4) are not related in general, in the sense that none of them imply the other.

Later, it was proved by Borg [3] a more general result: if

$$a \in L_T(\mathbf{R}, \mathbf{R}) \setminus \{0\}, \quad \int_0^T a(t)\, dt \geq 0, \quad \int_0^T |a(t)|\, dt \leq \frac{4}{T} \tag{1.5}$$

then (1.1) is stable.

Lyapunov's result is optimal in the following sense [3, 14]: for any $\varepsilon \in \mathbf{R}^+$, there are unstable differential equations (1.1) for which

$$a \in L_T(\mathbf{R}, \mathbf{R}) \setminus \{0\}, \quad \int_0^T a(t)\, dt \geq -\varepsilon, \quad \int_0^T |a(t)|\, dt \leq \frac{4}{T}$$

or for which

$$a \in L_T(\mathbf{R}, \mathbf{R}) \setminus \{0\}, \quad \int_0^T a(t)\, dt \geq 0, \quad \int_0^T |a(t)|\, dt \leq \frac{4}{T} + \varepsilon.$$

Condition (1.5) has been generalized in several ways For example, the authors provide in [18] optimal stability criteria by using L^p norms of a^+, $1 \leq p \leq \infty$, where a^+ is the positive part of the function a (i.e., $a^+(t) = \max\{0, a(t)\}$). See also [2, 12].

Returning to the paper [3], the proof given by Borg on the fact that (1.5) implies the stability of (1.1) is based on the following reasoning: if (1.1) is unstable, then, from Floquet theory, some characteristic multiplier s is real and different from zero. Consequently, there exists a nontrivial (not identically zero) solution $u = u(t)$ of (1.1) satisfying

$$u(t + T) = su(t), \quad \forall\, t \in \mathbf{R}. \tag{1.6}$$

Since $\int_0^T a(t)\,dt \geq 0$, the solution u must have some zero, and therefore, an infinity of zeros. From (1.6), it is deduced that the distance between two adjacent zeros t_1, t_2 does not exceed T.

Then, Borg writes that, due to a result by Beurling, the following inequality can be proved

$$\int_{t_1}^{t_2} |a(t)|\,dt > \frac{4}{t_2 - t_1}. \tag{1.7}$$

This is a clear contradiction with the hypothesis $\int_0^T |a(t)|\,dt \leq \frac{4}{T}$

We can conclude that the two problems:

(P1)   Stability of (1.1)
(P2)   The study of the existence of nontrivial solutions of the Dirichlet boundary value problem

$$u''(t) + a(t)u(t) = 0, \ t \in [t_1, t_2], \ u(t_1) = u(t_2) = 0$$

are strongly connected.

(P1) is also powerfully related to other boundary conditions, such as, for instance, Neumann, periodic, or antiperiodic boundary conditions (see [19]).

Hypotheses (1.3) and (1.4) can be seen as particular cases of a more general result. To see this, we introduce the parametric equation

$$u''(t) + (\mu + a(t))u(t) = 0, \ \mu \in \mathbf{R}. \tag{1.8}$$

Then, if $\lambda_n(a)$, $n \in \mathbf{N} \cup \{0\}$ and $\tilde{\lambda}_n(a)$, $n \in \mathbf{N}$, denote, respectively, the eigenvalues of (1.8) for the periodic ($u(0) - u(T) = u'(0) - u'(T) = 0$) and antiperiodic ($u(0) + u(T) = u'(0) + u'(T) = 0$) boundary conditions, Lyapunov and Haupt proved (see [8, 9, 14] for historical and mathematical details) that

$$\lambda_0(a) < \tilde{\lambda}_1(a) \leq \tilde{\lambda}_2(a) < \lambda_1(a) \leq \lambda_2(a) < \tilde{\lambda}_3(a) \leq \tilde{\lambda}_4(a) < \lambda_3(a) \leq \ldots \tag{1.9}$$

and that Eq. (1.8) is stable if

$$\mu \in (\lambda_{2n}(a), \tilde{\lambda}_{2n+1}(a)) \cup (\tilde{\lambda}_{2n+2}(a), \lambda_{2n+1}(a)) \tag{1.10}$$

for some $n \in \mathbf{N} \cup \{0\}$. As a consequence, if for some $n \in \mathbf{N} \cup \{0\}$, either

$$\lambda_{2n}(a) < 0 < \tilde{\lambda}_{2n+1}(a) \tag{1.11}$$

or

$$\tilde{\lambda}_{2n+2}(a) < 0 < \lambda_{2n+1}(a) \tag{1.12}$$

then (1.1) is stable. In this case, we say that $\mu = 0$ belongs to the nth stability zone of (1.8). In particular, (1.3) implies either (1.11) or (1.12) and the Lyapunov's conditions (1.4) imply $\lambda_0(a) < 0 < \tilde{\lambda}_1(a)$.

Except in very special cases, it is not an easy task to obtain some information on the sign of the previous eigenvalues, and it is at this point where the so-called Lyapunov inequalities can play an important role, as it is pointed out in this book (Sects. 3.3 and 5.5) and in the recent monograph [17].

We finish this section remarking that Lyapunov inequalities are usually referred to linear problems, but they can also be applied to the study of nonlinear resonant problems. This idea of using linear situations to obtain information on nonlinear ones is common in mathematics and it is present in many other situations: the linear Taylor polynomial, the implicit and inverse function theorems, Lyapunov stability with respect to the first approximation, etc. In the frame of Lyapunov inequalities, let us illustrate this with a simple example.

Consider the nonlinear Neumann problem

$$u''(x) + f(x, u(x)) = 0, \ x \in (0, L), \ u'(0) = u'(L) = 0, \tag{1.13}$$

where $f : [0, L] \times \mathbf{R} \to \mathbf{R}, \ (x, u) \to f(x, u)$, satisfies suitable regularity conditions [6, 11]. Let us rewrite (1.13) in the equivalent form

$$u''(x) + b(x, u(x))u(x) = -f(x, 0), \ x \in [0, L], \ u'(0) = u'(L) = 0, \tag{1.14}$$

where the function $b : [0, L] \times \mathbf{R} \to \mathbf{R}$ is defined by $b(x, z) = \int_0^1 f_u(x, \theta z) \, d\theta$.

Let us assume that we have a criterion, obtained from Lyapunov inequalities theory for linear equations, which implies the existence and uniqueness of solutions of linear problems like

$$u''(x) + a(x)u(x) = h(x), \ x \in [0, L], \ u'(0) = u'(L) = 0. \tag{1.15}$$

This criterion is usually given in terms of the function $a(\cdot)$. Now, assume that the function $b(\cdot, u(\cdot))$ fulfills this criterion for all function $u \in X = C^1([0, L], \mathbf{R})$.

Then, we can define the operator $T : X \to X$, by $Ty = u_y$ where u_y is the unique solution of the linear problem

$$u''(x) + b(x, y(x))u(x) = -f(x, 0), \ x \in [0, L], \ u'(0) = u'(L) = 0. \tag{1.16}$$

If we can define a norm on X such that T is completely continuous and bounded, the Schauder fixed point theorem provides a solution of (1.13). This is shown in this book in Sects. 2.2, 3.4, 4.3, 5.4, and 5.6.

Finally, we must say that although in this book we limit ourselves to the previously mentioned applications (stability of linear periodic equations, sign of the eigenvalues of eigenvalue problems, and nonlinear resonant problems), Lyapunov-type inequalities have a very broad range of applicability: crystallography, isoperimetric problems, Rayleigh type quotients, oscillation and intervals of disconjugacy, etc. Also, the theory has been generalized in different ways (higher-order problems, p-Laplacian operators, problems in Orlicz spaces, difference equations, ordinary differential equations with other terms of intermediate order derivatives, Hamiltonian systems, dynamic equations on time scales, etc.) The reader is referred, in a first instance, to [4, 5, 7, 10, 14, 17] for a more comprehensive understanding of the subject.

1.2   Overview of the Book

Next, we provide an overview of the contents of the book. To avoid unnecessary repetition, the bibliographic comments are made in the corresponding chapters.

Chapter 2 is devoted to present a general method for scalar ordinary differential equations with different boundary conditions (at the first eigenvalue in the case of resonant problems). This method allows to obtain the best Lyapunov constants from a variational point of view.

We begin with the definition and main properties of the L_p Lyapunov constant $(1 \leq p \leq \infty)$, in a given interval $(0, L)$, for Neumann boundary conditions. The variational characterization for these constants, as a minimum of some especial minimization problems defined in appropriate subsets X_p of the Sobolev space $H^1(0, L)$, is fundamental for several reasons: first, it allows to obtain an explicit expression for the L_p Lyapunov constant as a function of p and L; second, it allows the extension of the results to PDEs (Chap. 4) and to systems of equations (Chap. 5). For resonant problems (Neumann or periodic boundary conditions), it is necessary to impose an additional restriction to the definition of the spaces X_p, $1 \leq p \leq \infty$, so that we will have constrained minimization problems. This is not necessary in the case of nonresonant problems (Dirichlet or antiperiodic boundary conditions) where we will find unconstrained minimization problems. Moreover, the relation between Neumann boundary conditions and disfocality is considered. In addition, by using the Schauder fixed point theorem, we apply the obtained results for linear equations to nonlinear problems.

In Chap. 3 we study L_1 Lyapunov-type inequalities for different boundary conditions at higher eigenvalues.

A key point is a detailed analysis about the number and distribution of zeros of nontrivial solutions and their first derivatives. To this respect, we compare our problem with other one with mixed boundary conditions. This allows us to obtain a more precise information than that obtained when the classical Sturm separation theorem is used, where the problem is compared with the case of Dirichlet boundary conditions.

After this, we consider different suitable minimization problems, which allow to obtain the best L_1 Lyapunov constant at higher eigenvalues. As in the classical result by Lyapunov at the first eigenvalue, the L_1 best constant is not attained. The linear study on periodic and antiperiodic boundary conditions is used to establish some new conditions for the stability of linear periodic equations, like the Hill's equation. The notion of disfocality at higher eigenvalues is briefly discussed and, again, the Schauder fixed point theorem is used to provide sufficient conditions about the existence and uniqueness of solutions for resonant nonlinear problems at higher eigenvalues.

Chapter 4 concerns the study of L_p Lyapunov inequalities for partial differential equations in a regular and bounded domain $\Omega \subset \mathbf{R}^N$. It begins with the study of linear problems with Neumann boundary conditions. Significant differences are shown from the ordinary case, which explain, for instance, the sentence: L_1 *Lyapunov inequality is meaningless for PDEs*.

Perhaps, the most important difference is, without doubt, the following fact: if $N = 2$, there is L_p Lyapunov inequality if and only if $1 < p \leq \infty$, and if $N \geq 3$, there is L_p Lyapunov inequality if and only if $N/2 \leq p \leq \infty$. In particular, we must remark (again) that there is no L_1 Lyapunov inequality in the PDE case and that the best constant is attained if $N/2 < p \leq \infty$. The shown variational method makes possible to obtain analogous results in the case of Dirichlet boundary conditions. As in the ordinary case, we apply the obtained result to nonlinear resonant problems.

The chapter ends with the treatment of radial higher eigenvalues. We treat the case of Neumann boundary conditions on balls in $\mathbf{R}^N$. By using appropriate minimizing sequences and a detailed analysis about the number and distribution of zeros of radial nontrivial solutions, we show significant qualitative differences according to the studied case is subcritical, supercritical or critical. Some facts related to Sturm type comparison results are fundamental in the proof.

Chapter 5 deals with the study of L_p Lyapunov-type inequalities for linear systems of equations with different boundary conditions (which include the case of Neumann, Dirichlet, periodic, and antiperiodic boundary value problems) and for any constant $p \geq 1$. Previously, we explain, from our point of view, the meaning of the sentence "Lyapunov inequalities for systems of equations."

Ordinary and elliptic problems are considered. Resonant nonlinear problems as well as the stable boundedness of linear periodic conservative systems are also studied. The proof uses again, in a fundamental way, the nontrivial relation (proved in Chap. 2 for ODEs and in Chap. 4 for PDEs) between the best Lyapunov constants and the minimum value of some especial (constrained or unconstrained) minimization problems.

References

1. Berger, M.S.: Nonlinearity and Functional Analysis: Lectures on Nonlinear Problems in Mathematical Analysis. Academic, New York (1977)
2. Borg, G.: Über die Stabilität gewisser Klassen von linearen Differentialgleichungen. Ark. Mat. Astr. Fys. **31** A(1), 1–31 (1944)

3. Borg, G.: On a Liapounoff criterion of stability. Am. J. Math. **71**, 67–70 (1949)
4. Brown, R.C., Hinton, D.B.: Lyapunov inequalities and their applications. In: Survey on Classical Inequalities. Mathematics and Its Applications, vol. 517, pp. 1–25. Kluwer Academic, Dordrecht (2000)
5. Cañada, A., Villegas, S.: An applied mathematical excursion through Lyapunov inequalities, classical analysis and differential equations. SeMa J. Soc. Esp. de Matemática Apl. **57**, 69–106 (2012)
6. Cañada, A., Montero, J.A., Villegas, S. : Lyapunov type inequalities and Neumann boundary value problems at resonance. Math. Inequal. Appl. **8**, 459–475 (2005)
7. Cheng, S.: Lyapunov inequalities for differential and difference equations. Fasc. Math. **23**, 25–41 (1992)
8. Coddington, E.A., Levinson, N.: Theory of Ordinary Differential Equations. McGraw-Hill, New York (1955)
9. Hale, J.K.: Ordinary Differential Equations. Wiley-Interscience (Wiley), New York (1969)
10. Hartman, P.: Ordinary Differential Equations. Wiley, New York (1964)
11. Huaizhong, W., Yong, L.: Neumann boundary value problems for second-order ordinary differential equations across resonance. SIAM J. Control. Optim. **33**, 1312–1325 (1995)
12. Krein, M.G.: On Certain Problems on the Maximum and Minimum of Characteristic Values and on the Lyapunov Zones of Stability. American Mathematical Society Translations, Series 2, vol. 1. American Mathematical Society, Providence, RI (1955)
13. Lyapunov, M.A.: Problème général de la stabilité du mouvement. Ann. Fac. Sci. Univ. Tolouse Sci. Math. Sci. Phys. **9**, 203–474 (1907)
14. Magnus, W., Winkler, S.: Hill's Equation. Dover, New York (1979)
15. Mawhin, J., Ward, J.R.: Nonuniform nonresonance conditions at the two first eigenvalues for periodic solutions of forced Liénard and Duffing equations. Rocky Mt. J. Math. **12**, 643–654 (1982)
16. Mawhin, J., Ward, J.R.: Periodic solutions of some forced Liénard differential equations at resonance. Arch. Math. (Basel) **41**, 337–351 (1983)
17. Pinasco, J.P.: Lyapunov-Type Inequalities. With Applications to Eigenvalue Problems. Springer Briefs in Mathematics. Springer, New York (2013)
18. Zhang, M., Li, W.: A Lyapunov-type stability criterion using L^α norms. Proc. Am. Math. Soc. **130**, 3325–3333 (2002)
19. Zhang, M.: Certain classes of potentials for p-Laplacian to be non-degenerate. Math. Nachr. **278**, 1823–1836 (2005)

Chapter 2
A Variational Characterization of the Best Lyapunov Constants

Abstract This chapter is devoted to the definition and main properties of the L_p Lyapunov constant, $1 \le p \le \infty$, for scalar ordinary differential equations with different boundary conditions, in a given interval $(0, L)$. It includes resonant problems at the first eigenvalue and nonresonant problems. A main point is the characterization of such a constant as a minimum of some especial minimization problem, defined in appropriate subsets X_p of the Sobolev space $H^1(0, L)$. This variational characterization is a fundamental fact for several reasons: first, it allows to obtain an explicit expression for the L_p Lyapunov constant as a function of p and L; second, it allows the extension of the results to systems of equations (Chap. 5) and to PDEs (Chap. 4). For resonant problems (Neumann or periodic boundary conditions), it is necessary to impose an additional restriction to the definition of the spaces X_p, $1 \le p \le \infty$, so that we will have constrained minimization problems. This is not necessary in the case of nonresonant problems (Dirichlet or antiperiodic boundary conditions) where we will find unconstrained minimization problems. For nonlinear equations, we combine the Schauder fixed point theorem with the obtained results for linear equations.

2.1 Neumann Boundary Conditions, As a Paradigm of Linear Resonant Problems

This section will be concerned with the existence of nontrivial solutions of the homogeneous linear problem with Neumann boundary conditions

$$u''(x) + a(x)u(x) = 0, \; x \in (0, L), \; u'(0) = u'(L) = 0. \qquad (2.1)$$

If $a(\cdot)$ is a constant function $\lambda \in \mathbf{R}$, (2.1) has nontrivial solutions if and only if λ belongs to the set $\{\lambda_n = n^2\pi^2/L^2, \; n \in \mathbf{N} \cup \{0\}\}$, i.e., the set of eigenvalues of the eigenvalue problem:

$$u''(x) + \lambda u(x) = 0, \; x \in (0, L), \; u'(0) = u'(L) = 0. \qquad (2.2)$$

Obviously, the problem is much more complicated if $a(\cdot)$ is not a constant function.

© The Author(s) 2015
A. Cañada, S. Villegas, *A Variational Approach to Lyapunov Type Inequalities*,
SpringerBriefs in Mathematics, DOI 10.1007/978-3-319-25289-6_2

To proceed to the definition of the Lyapunov constants, we will assume through-out this chapter that $a \in \Lambda$, where Λ is defined by

$$\Lambda = \{a \in L^1(0,L) \setminus \{0\} : \int_0^L a(x)\, dx \geq 0 \text{ and } (2.1) \text{ has nontrivial solutions }\}.$$

(2.3)

Here, for each p, $1 \leq p < \infty$, $L^p(0,L)$ denotes the usual Lebesgue space of measurable functions $a(\cdot)$ such that $|a(\cdot)|^p$ is integrable in $(0,L)$, while $L^\infty(0,L)$ denotes the set of measurable functions such that there exists a constant c satisfying $|a(x)| \leq c$, a.e. in $(0,L)$. On the other hand, $u \in H^1(0,L)$, the usual Sobolev space. The interested reader can consult the reference [2] for these concepts.

For each p with $1 \leq p \leq \infty$, we can define the functional $I_p : \Lambda \bigcap L^p(0,L) \rightarrow \mathbf{R}$ given by the expression

$$I_p(a) = \|a^+\|_p = \left(\int_0^L |a^+(x)|^p\, dx\right)^{1/p}, \forall\, a \in \Lambda \bigcap L^p(0,L),\ 1 \leq p < \infty$$

$$I_\infty(a) = \|a^+\|_\infty = \sup\ \text{ess}\ a^+, \forall a \in \Lambda \bigcap L^\infty(0,L),$$

(2.4)

where a^+ is the positive part of the function a (i.e., $a^+(x) = \max\{0, a(x)\}$) and sup ess a^+ is the essential supremum of the function a^+.

Since the positive eigenvalues of the eigenvalue problem (2.2), belong to the set $\Lambda \bigcap L^p(0,L)$, the nonnegative constant

$$\beta_p \equiv \inf_{a \in \Lambda \bigcap L^p(0,L)} I_p(a),\ 1 \leq p \leq \infty$$

(2.5)

is well defined. Due to the pioneering work of Lyapunov for Dirichlet boundary conditions and $p = 1$ [16, 22, 23], we will call to the constant β_p, defined in (2.5), the *best (optimal) L_p Lyapunov constant*.

Remark 2.1. We need the positivity of $\int_0^L a(x)\, dx$ in order to prove that the constant β_p is strictly positive. In fact, if the set Λ in (2.3) is replaced by

$$\Upsilon = \{a \in L^1(0,L) \setminus \{0\} : (2.1) \text{ has nontrivial solutions }\}$$

then the constant

$$\gamma_p \equiv \inf_{a \in \Upsilon \bigcap L^p(0,L)} I_p(a),\ 1 \leq p \leq \infty$$

is zero, for each p, $1 \leq p \leq \infty$ (see Remark 4 in [3]). The real number 0 is the first eigenvalue of the eigenvalue problem (2.2). As it will be seen in Sect. 2.3, this *extra condition* on the sign of $\int_0^L a(x)\, dx$ is not necessary in nonresonant problems.

Remark 2.2. The study of the constant β_p can be seen as an optimal control problem: the admissible control set is $\Lambda \cap L^p(0, L)$ and the functional that we want to minimize is I_p. However, we caution that the condition

$$(2.1) \text{ has nontrivial solutions} \tag{2.6}$$

is difficult to handle from a mathematical point of view and this is the main difficulty of the problem. Because of this, one of the main purposes of this chapter is to get a variational characterization of the best Lyapunov constant β_p. This will be very important for the possible extension of the results to PDEs and to systems of equations.

We begin with the easiest situation: $p = \infty$. In this case, the constant β_∞ is nothing but the first positive eigenvalue of (2.2). The proof is known and it uses two basic ideas: Hölder's inequality and the variational characterization of the eigenvalues of (2.2) [8].

Theorem 2.1.

$$\beta_\infty = \min_{v \in X_\infty \setminus \{0\}} \frac{\displaystyle\int_0^L (v')^2}{\displaystyle\int_0^L (v)^2} = \frac{\pi^2}{L^2}, \tag{2.7}$$

where $X_\infty = \{v \in H^1(0, L) : \displaystyle\int_0^L v = 0\}$.

Proof. If $a \in \Lambda$ and $u \in H^1(0, L)$ is a nontrivial solution of (2.1), then

$$\int_0^L u'v' = \int_0^L auv, \ \forall \ v \in H^1(0, L).$$

In particular, we have

$$\int_0^L u'^2 = \int_0^L au^2, \ \int_0^L au = 0. \tag{2.8}$$

Therefore, for each $k \in \mathbf{R}$, we have

$$\int_0^L (u+k)'^2 = \int_0^L u'^2 = \int_0^L au^2 \leq \int_0^L au^2 + k^2 \int_0^L a$$

$$= \int_0^L au^2 + \int_0^L k^2 a + 2k \int_0^L au = \int_0^L a(u+k)^2 \leq \int_0^L a^+(u+k)^2.$$

Hölder's inequality implies

$$\int_0^L (u+k)'^2 \le \|a^+\|_\infty \int_0^L (u+k)^2.$$

Also, since the function a belongs to Λ, u is a nonconstant solution of (2.1), so that $u + k$ is a nontrivial function. Consequently

$$\|a^+\|_\infty \ge \frac{\displaystyle\int_0^L (u+k)'^2}{\displaystyle\int_0^L (u+k)^2}.$$

Now, choose $k_0 \in \mathbf{R}$ satisfying

$$\int_0^L (u+k_0) = 0. \tag{2.9}$$

Then,

$$\|a^+\|_\infty \ge \frac{\displaystyle\int_0^L (u+k_0)'^2}{\displaystyle\int_0^L (u+k_0)^2} \ge \inf_{v \in X_\infty \setminus \{0\}} \frac{\displaystyle\int_0^L (v')^2}{\displaystyle\int_0^L (v)^2} = \frac{\pi^2}{L^2}, \; \forall\, a \in \Lambda. \tag{2.10}$$

Moreover, it is very well known that the previous infimum is, in fact, a minimum and that the value of this minimum is $\frac{\pi^2}{L^2}$ [8]. The previous inequalities imply $\beta_\infty \ge \frac{\pi^2}{L^2}$. Since the constant function $\frac{\pi^2}{L^2}$ is an element of Λ, we deduce $\beta_\infty = \frac{\pi^2}{L^2}$. This completes the proof of the theorem.

Remark 2.3. The constant β_∞ was defined in (2.5) as an infimum, but it can be seen that this infimum is attained in a unique element $a_\infty \in \Lambda$, given by $a_\infty(x) \equiv \frac{\pi^2}{L^2}$ [3].

Now we deal with the case $p = 1$. It is the only case where the infimum β_p, defined in (2.5), is not attained. The proof is inspired by Borg [1], but next theorem additionally provides a variational characterization of β_1 [3].

Theorem 2.2.

$$\beta_1 = \min_{u \in X_1 \setminus \{0\}} \frac{\displaystyle\int_0^L u'^2}{\|u\|_\infty^2} = \frac{4}{L}, \tag{2.11}$$

where $X_1 = \{u \in H^1(0,L) : \max_{x \in [0,L]} u(x) + \min_{x \in [0,L]} u(x) = 0\}.$

Proof. First, we prove

$$\min_{u \in X_1 \setminus \{0\}} \frac{\int_0^L u'^2}{\|u\|_\infty^2} = \frac{4}{L}. \tag{2.12}$$

To do this, if $u \in X_1 \setminus \{0\}$, and $x_1, x_2 \in [0, L]$ are such that $u(x_1) = \max_{[0,L]} u$, $u(x_2) = \min_{[0,L]} u$, then $\|u\|_\infty = \max_{[0,L]} u = -\min_{[0,L]} u$. Clearly, it is not restrictive to assume that $x_1 < x_2$. Let us denote $I = [x_1, x_2]$. Then, it follows from the Cauchy–Schwarz inequality

$$\int_0^L u'^2 \geq \int_I u'^2 \geq \frac{\left(\int_I |u'|\right)^2}{x_2 - x_1} \geq \frac{\left(\int_I u'\right)^2}{x_2 - x_1}$$
$$= \frac{(u(x_2) - u(x_1))^2}{x_2 - x_1} = \frac{4\|u\|_\infty^2}{x_2 - x_1} \geq \frac{4}{L}\|u\|_\infty^2. \tag{2.13}$$

Therefore,

$$\inf_{u \in X_1 \setminus \{0\}} \frac{\int_0^L u'^2}{\|u\|_\infty^2} \geq \frac{4}{L}.$$

On the other hand, if $v(x) = x - \frac{L}{2}$, $\forall x \in [0, L]$, then $v \in X_1 \setminus \{0\}$ and $\dfrac{\int_0^L v'^2}{\|v\|_\infty^2} = \dfrac{4}{L}$. This proves (2.12).

Now, we prove that $\beta_1 = \dfrac{4}{L}$. To see this, if $a \in \Lambda$ and $u \in H^1(0, L)$ is a nontrivial solution of (2.1), then by using Hölder's inequality, we obtain for each $k \in \mathbf{R}$,

$$\int_0^L (u + k)'^2 \leq \int_0^L a(u + k)^2 \leq \|a^+\|_1 \|(u + k)\|_\infty^2$$

and consequently

$$\|a^+\|_1 \geq \frac{\int_0^L (u + k)'^2}{\|(u + k)\|_\infty^2}.$$

If we choose $k_0 \in \mathbf{R}$ satisfying $u + k_0 \in X_1$, we deduce

$$\|a^+\|_1 \geq \frac{\int_0^L (u + k_0)'^2}{\|u + k_0\|_\infty^2} \geq \frac{4}{L}, \quad \forall\, a \in \Lambda. \tag{2.14}$$

Therefore, $\beta_1 \geq \dfrac{4}{L}$. Also, we can define a minimizing sequence in the following way. Let $\{u_n\} \subset C^2[0, L]$ be a sequence such that $u_n(x) = \left(x - \frac{L}{2}\right)$, $\forall\, x \in \left(\frac{1}{n}, L - \frac{1}{n}\right)$; $u_n'(0) = u_n'(L) = 0$; $u_n''(x) > 0$, $\forall\, x \in \left[0, \frac{1}{n}\right)$; $u_n''(x) < 0$, $\forall\, x \in \left(L - \frac{1}{n}, L\right]$. Then, if we define the sequence of continuous functions $a_n : [0, L] \to \mathbf{R}$, as $a_n(x) = 0$, $\forall\, x \in \left[\frac{1}{n}, L - \frac{1}{n}\right]$; $a_n(x) = \dfrac{-u_n''(x)}{u_n(x)}$, $\forall\, x \in \left[0, \frac{1}{n}\right] \cup \left[L - \frac{1}{n}, L\right]$, we have that $a_n \in L^\infty(0, L)$, $a_n \geq 0$, a.e. in $(0, L)$, a_n is nontrivial and

$$u_n''(x) + a_n(x) u_n(x) = 0, \ \text{in } (0, L), \ u_n'(0) = u_n'(L) = 0.$$

Therefore, $a_n \in \Lambda$, $\forall\, n \in \mathbf{N}$. Moreover,

$$\int_0^L a_n^+ = \int_0^{\frac{1}{n}} \frac{-u_n''(x)}{u_n(x)} + \int_{L-\frac{1}{n}}^L \frac{-u_n''(x)}{u_n(x)}$$

$$\leq \int_0^{\frac{1}{n}} \frac{u_n''(x)}{\min\limits_{[0,\frac{1}{n}]} (-u_n)} + \int_{L-\frac{1}{n}}^L \frac{-u_n''(x)}{\min\limits_{[L-\frac{1}{n},L]} (u_n)}$$

$$= \frac{u_n'\left(\frac{1}{n}\right)}{\frac{L}{2} - \frac{1}{n}} + \frac{u_n'\left(L - \frac{1}{n}\right)}{\frac{L}{2} - \frac{1}{n}} = \frac{1}{\frac{L}{2} - \frac{1}{n}} + \frac{1}{\frac{L}{2} - \frac{1}{n}}.$$

Taking limits as $n \to \infty$, we deduce $\beta_1 = \dfrac{4}{L}$.

Remark 2.4. The infimum β_1, defined in (2.5), is not attained, i.e.,

$$\|a^+\|_1 > \frac{4}{L}, \ \forall\, a \in \Lambda. \tag{2.15}$$

To prove this, let $a \in \Lambda$ be such that $\|a^+\|_1 = \dfrac{4}{L}$. By choosing u a nontrivial solution of (2.1) and $k_0 \in \mathbf{R}$ such that $u + k_0 \in X_1$, we obtain

$$\int_0^L (u + k_0)'^2 \leq \frac{4}{L} \|(u + k_0)\|_\infty^2.$$

On the other hand, since $u + k_0 \in X_1$, we deduce from (2.12)

$$\int_0^L (u + k_0)'^2 \geq \frac{4}{L} \|(u + k_0)\|_\infty^2.$$

Therefore,

$$\int_0^L (u + k_0)'^2 = \frac{4}{L} \|(u + k_0)\|_\infty^2.$$

Then, for the function $u + k_0$, all the inequalities of (2.13) transform into equalities. In particular, $x_2 = L$, $x_1 = 0$ and $\left(\int_0^L (u + k_0)' \right)^2 = L \int_0^L (u + k_0)'^2$. Again, the Cauchy–Schwarz inequality (equality in this case) implies that the function $(u + k_0)'$ is constant in $[0, L]$. Taking into account that $u + k_0 \in X_1 \setminus \{0\}$, we have $u(x) + k_0 = k(x - \frac{L}{2})$, $\forall x \in [0, L]$ and for some nontrivial constant k. Then from (2.1) we deduce $a \equiv 0$, which is a contradiction.

Remark 2.5. The formula $\beta_1 = \dfrac{4}{L}$ was proved in [18] by using methods from Optimal Control Theory. More precisely, the authors used the Pontryagin's maximum principle. The variational proof that we have presented here motivates some of the main ideas that we will use in the case $1 < p < \infty$.

Remark 2.6. In [21] the authors study the problem with linear damping

$$u''(x) + b(x)u'(x) + a(x)u(x) = 0, \ u'(0) = u'(L) = 0 \tag{2.16}$$

obtaining the best L_1 Lyapunov constant.

As a first application of Theorems 2.1 and 2.2 to the linear problem

$$u''(x) + a(x)u(x) = f(x), \ x \in (0, L), \ u'(0) = u'(L) = 0 \tag{2.17}$$

we have the following corollary.

Corollary 2.1. *Let $a \in L^\infty \setminus \{0\}$, $0 \leq \displaystyle\int_0^L a(x)$, satisfying one of the following conditions:*

1. $\|a^+\|_1 \leq \beta_1 = \dfrac{4}{L}$

2. $\|a^+\|_\infty \leq \beta_\infty = \dfrac{\pi^2}{L^2}$ and a^+ is not identically to the constant β_∞

Then for each $f \in L^\infty(0, L)$, the boundary value problem (2.17) has a unique solution.

Proof. The corollary is proved if the homogeneous problem

$$u''(x) + a(x)u(x) = 0, \ x \in (0, L), \ u'(0) = u'(L) = 0 \tag{2.18}$$

has only the trivial solution [16]. But this is clear from Theorem 2.1, Remark 2.3 and Theorem 2.2, Remark 2.4.

Remark 2.7. In the previous corollary, the conditions on the function $a(\cdot)$:

$$\begin{aligned} \|a^+\|_1 &\leq \beta_1 \\ \|a^+\|_\infty \leq \beta_\infty = \ &\text{and } a^+ \text{ is not identically to the constant } \beta_\infty \end{aligned} \tag{2.19}$$

are given, respectively, in terms of the L^1 norm $\|a^+\|_1$ and L^∞ norm $\|a^+\|_\infty$. Clearly, they are not related in general, in the sense that none of them imply the other. In the next theorem, we consider the case $1 < p < \infty$, and we establish other different conditions given in terms of the L^p norm $\|a^+\|_p$, $1 < p < \infty$. They will show a natural relation between the cases $p = 1$ and $p = \infty$ in (2.19) when one studies what happens for $p \to 1^+$ and $p \to \infty$.

In order to motivate the variational characterization of the constant β_p, $1 < p < \infty$, which is discussed in the next theorem, take into account that if $a \in \Lambda \bigcap L^p(0, L)$ and $u \in H^1(0, L)$ is a nontrivial solution of (2.1) then

$$\int_0^L u'v' = \int_0^L auv, \ \forall \, v \in H^1(0, L).$$

In particular, choosing $v \equiv u$ and $v \equiv 1$, we have respectively

$$\int_0^L u'^2 = \int_0^L au^2, \ \int_0^L au = 0. \tag{2.20}$$

Therefore, for each $k \in \mathbf{R}$, we have (remember that $\int_0^L a \geq 0$)

$$\int_0^L (u+k)'^2 = \int_0^L u'^2 = \int_0^L au^2 \leq \int_0^L au^2 + k^2 \int_0^L a$$

$$= \int_0^L au^2 + \int_0^L k^2 a + 2k \int_0^L au = \int_0^L a(u+k)^2 \leq \int_0^L a^+(u+k)^2.$$

From Hölder's inequality it follows

$$\int_0^L (u+k)'^2 \leq \|a^+\|_p \|(u+k)^2\|_{\frac{p}{p-1}}.$$

Moreover, since u is a nonconstant solution of (2.1), $u+k$ is not identically the zero function. Consequently

$$\|a^+\|_p \geq \frac{\displaystyle\int_0^L (u+k)'^2}{\|(u+k)^2\|_{\frac{p}{p-1}}}, \ \forall \, a \in \Lambda. \tag{2.21}$$

This reasoning suggests the minimization of a functional like the previous one on some *appropriate subset* of $H^1(0, L)$. Motivated by the case $p = \infty$ (Theorem 2.1), this *appropriate subset* could be of the type

$$\left\{ u \in H^1(0, L) : \int_0^L |u|^{\lambda(p)} u = 0 \right\},$$

where $\lambda(\infty) = 0$. Here we choose $\lambda(p) = \dfrac{2}{p-1}$. To understand why this election is suitable, we must see in detail the proof of the next theorem, especially the part where the Lagrange multiplier Theorem is applied (see [3] for more details).

Theorem 2.3. *If* $1 < p < \infty$,

$$\beta_p = \min_{X_p\setminus\{0\}} J_p(u) = \min_{X_p\setminus\{0\}} \frac{\displaystyle\int_0^L u'^2}{\left(\displaystyle\int_0^L |u|^{\frac{2p}{p-1}}\right)^{\frac{p-1}{p}}}$$

$$= \frac{4(p-1)^{1+\frac{1}{p}}}{L^{2-\frac{1}{p}}p(2p-1)^{1/p}}\left(\int_0^{\pi/2}(\sin x)^{-1/p}\,dx\right)^2, \qquad (2.22)$$

where

$$X_p = \left\{u \in H^1(0,L) : \int_0^L |u|^{\frac{2}{p-1}}u = 0\right\}.$$

Proof. The proof will be carried out into three steps:

1. **The minimization problem stated in (2.22) has solution.**

   The proof of this fact is standard: first we will demonstrate that any minimizing sequence is bounded in the Hilbert space $H^1(0,L)$. Then we will use that the considered functional is weak lower semi-continuous in order to conclude that the infimum is attained.

   Let us denote

$$m_p \equiv \inf_{X_p\setminus\{0\}} J_p. \qquad (2.23)$$

If $\{u_n\} \subset X_p \setminus \{0\}$ is a minimizing sequence, then $\{k_n u_n\}$ where $\{k_n\}$ is an arbitrary sequence of nonzero real numbers, is also a minimizing sequence, since $J_p(u_n) = J_p(k_n u_n)$. Therefore, we can assume without loss of generality that $\int_0^L |u_n|^{\frac{2p}{p-1}} = 1$. As $J_p(u_n)$ is bounded, $\left\{\int_0^L |u_n'^2|\right\}$ is also bounded. Moreover, since $\int_0^L |u_n|^{\frac{2}{p-1}}u_n = 0$, for each u_n there is $x_n \in (0,L)$ such that $u_n(x_n) = 0$. Now, $u_n(x) = \int_{x_n}^x u'(s)\,ds$, $\forall x \in (0,L)$ and Hölder's inequality implies that $\{u_n\}$ is bounded in $H^1(0,L)$. So, we can suppose, up to a subsequence, that $u_n \rightharpoonup u_0$ in $H^1(0,L)$ (weak convergence) and $u_n \to u_0$ in $C[0,L]$, with the uniform norm [2]. The strong convergence in $C[0,L]$ gives us $\int_0^L |u_0|^{\frac{2p}{p-1}} = 1$, $\int_0^L |u_0|^{\frac{2}{p-1}}u_0 = 0$, and consequently $u_0 \in X_p \setminus \{0\}$. As the functional J_p is weak lower semi-continuous [2], the weak convergence in $H^1(0, L)$ implies $J_p(u_0) \leq \liminf J_p(u_n) = m_p$. Then u_0 is a minimizer.

Since $X_p = \{ u \in H^1(0, L) : \varphi(u) = 0 \}$, $\varphi(u) = \int_0^L |u|^{\frac{2}{p-1}} u$, if $u_0 \in X_p \setminus \{0\}$ is any minimizer of J_p, Lagrange multiplier Theorem [10] implies that there is $\lambda \in \mathbf{R}$ such that

$$H'(u_0) + \lambda \varphi'(u_0) = 0,$$

where $H : H^1(0, L) \to \mathbf{R}$ is defined by

$$H(u) = \int_0^L u'^2 - m_p \left(\int_0^L |u|^{\frac{2p}{p-1}} \right)^{\frac{p-1}{p}}.$$

Also, as $u_0 \in X_p$ we have $H'(u_0)(1) = 0$. Moreover $H'(u_0)(v) = 0$, $\forall\, v \in H^1(0, L) : \varphi'(u_0)(v) = 0$. Finally, as any $v \in H^1(0, L)$ may be written in the form $v = \alpha + w$, $\alpha \in \mathbf{R}$, and w satisfying $\varphi'(u_0)(w) = 0$, we conclude $H'(u_0)(v) = 0$, $\forall\, v \in H^1(0, L)$, i.e., $H'(u_0) = 0$ which implies that u_0 satisfies the problem

$$v''(x) + m_p \left(\int_0^L |v|^{\frac{2p}{p-1}} \right)^{\frac{-1}{p}} |v(x)|^{\frac{2}{p-1}} v(x) = 0, \quad v'(0) = v'(L) = 0. \quad (2.24)$$

2. **The constant β_p is equal to the constant m_p** (this fact implies the characterization of β_p as the minimum value of J_p on $X_p \setminus \{0\}$ and will be of special interest in the extension of the results to systems of equations in Chap. 5).

In fact, previously to the theorem, we have proved that if $a \in \Lambda \bigcap L^p(0, L)$ and $u \in H^1(0, L)$ is a nontrivial solution of (2.1), then (2.21) is satisfied for each $k \in \mathbf{R}$. Then, if for each $a \in \Lambda \bigcap L^p(0, L)$ and each u, nontrivial solution of (2.1), we choose $k_0 \in \mathbf{R}$ satisfying $u + k_0 \in X_p$, we deduce $\beta_p \geq m_p$. Reciprocally, if $u_p \in X_p \setminus \{0\}$ is any minimizer of J_p, then u_p satisfies (2.24). Therefore, if we denote

$$A_p(v) = m_p \left(\int_0^L |v|^{\frac{2p}{p-1}} \right)^{\frac{-1}{p}} \quad (2.25)$$

we have that $A_p(u_p)|u_p|^{\frac{2}{p-1}} \in \Lambda \bigcap L^p(0, L)$ and

$$\| A_p(u_p)|u_p|^{\frac{2}{p-1}} \|_p = m_p.$$

Then $\beta_p \leq m_p$. The conclusion is that $\beta_p = m_p$.

3. **Integrating the Euler's equation (2.24) to obtain m_p.**

The explicit calculus of m_p is a very delicate and technical matter, but we emphasize that the same ideas can be used in many other situations, as it will be seen in Sect. 2.3 (see [3] for further details). In fact, this method can be used whenever we have a detailed knowledge about the number and distribution of zeros of nontrivial solutions v of Eq. (2.26) below and their first derivatives v'.

Start with the method: if $u_p \in X_p \setminus \{0\}$ is a minimizer of J_p, then we have proved that u_p satisfies a problem of the type

$$v''(x) + B|v(x)|^{\frac{2}{p-1}} v(x) = 0, \ x \in, (0, L), \ v'(0) = v'(L) = 0, \qquad (2.26)$$

where B is some positive real constant. Also, let us observe that if v is a nontrivial solution of (2.26), then $\int_0^L |v(x)|^{\frac{2}{p-1}} v(x) = 0$. Therefore, v belongs to $X_p \setminus \{0\}$ and consequently,

$$\inf_{B \in \mathbf{R}^+} \ \inf_{v \in S_B} J_p(v) = m_p,$$

where, for a given $B \in \mathbf{R}^+$, S_B denotes the set of all nontrivial solutions of (2.26).

Now, let $B \in \mathbf{R}^+$ be a fixed number and v a nontrivial solution of (2.26). First, our main purpose is *to calculate* v in the interval $[0, L]$ and then, *to calculate* $J_p(v)$. It is clear that we may assume without loss of generality that $v(0) > 0$. Moreover, since $v \in X_p \setminus \{0\}$, v must change its sign in $(0, L)$. Let x_0 be the first zero point of v in $(0, L)$.

a. The function v in $[0, x_0]$.

The function v satisfies the initial value problem

$$w''(x) + B|w(x)|^{\frac{2}{p-1}} w(x) = 0, \ w(0) = v(0), \ w'(0) = 0 \qquad (2.27)$$

and this problem has a unique solution defined in $\mathbf{R}$ (see Proposition 2.1. in [14]).

If $x \in (0, x_0)$ is fixed, multiplying both terms of (2.26) by v' and integrating in the interval $[0, x]$ we obtain

$$-\frac{(v'(x))^2}{2} = \frac{B(p-1)}{2p} \left(|v(x)|^{\frac{2p}{p-1}} - |v(0)|^{\frac{2p}{p-1}} \right). \qquad (2.28)$$

On the interval $(0, x_0)$ the function v satisfies $v(x) > 0$ and $v'(x) \le 0$ (see (2.26)) and thus

$$v'(x) = -\left[\frac{B(p-1)}{p} \right]^{1/2} \left[|v(0)|^{\frac{2p}{p-1}} - |v(x)|^{\frac{2p}{p-1}} \right]^{1/2}. \qquad (2.29)$$

Therefore,

$$\int_0^x \frac{v'(t)}{\left[|v(0)|^{\frac{2p}{p-1}} - |v(t)|^{\frac{2p}{p-1}}\right]^{1/2}} \, dt = -\left[\frac{B(p-1)}{p}\right]^{1/2} x$$

for any $x \in (0, x_0)$. Doing the change of variables $s = \dfrac{v(t)}{v(0)}$, previous relation can be written as

$$-\varphi(1) + \varphi\left(\frac{v(x)}{v(0)}\right) = -v(0)^{\frac{1}{p-1}}\left[\frac{B(p-1)}{p}\right]^{1/2} x, \quad \forall \, x \in (0, x_0).$$

Here $\varphi : [0, 1] \to \mathbf{R}$ is the strictly increasing function defined by

$$\varphi(t) = \int_0^t \frac{ds}{\left(1 - s^{\frac{2p}{p-1}}\right)^{1/2}}.$$

If $\varphi\,[0, 1] = [0, I]$, then we find

$$\frac{v(x)}{v(0)} = \varphi^{-1}\left[I - v(0)^{\frac{1}{p-1}}\left(\frac{B(p-1)}{p}\right)^{1/2} x\right] \quad \forall \, x \in (0, x_0). \tag{2.30}$$

Moreover, since $v(x_0) = 0$, we obtain

$$I - v(0)^{\frac{1}{p-1}}\left(\frac{B(p-1)}{p}\right)^{1/2} x_0 = 0.$$

Hence,

$$v(0) = \left(\frac{I}{x_0}\left(\frac{p}{B(p-1)}\right)^{1/2}\right)^{p-1}. \tag{2.31}$$

Finally,

$$v(x) = \left(\frac{I}{x_0}\left(\frac{p}{B(p-1)}\right)^{1/2}\right)^{p-1} \varphi^{-1}\left(I - \frac{I}{x_0}x\right), \quad \forall \, x \in [0, x_0] \tag{2.32}$$

b. Now, we can calculate v in $[x_0, 2x_0], [2x_0, 3x_0], \ldots$
   To do this, the initial value problem

$$w''(x) + B|w(x)|^{\frac{2}{p-1}} w(x) = 0, \quad w(x_0) = v(x_0) = 0, \quad w'(x_0) = v'(x_0) \tag{2.33}$$

has a unique solution defined in **R** [14]. Since the function $-v(2x_0 - x)$, $x \in (x_0, 2x_0)$, is a solution of (2.33), this provides $v(x) = -v(2x_0 - x)$, $\forall x \in (x_0, 2x_0)$.

In an analogous way, the initial value problem

$$w''(x) + B|w(x)|^{\frac{2}{p-1}} w(x) = 0, \quad w(2x_0) = v(2x_0), \quad w'(2x_0) = v'(2x_0) = 0$$
(2.34)

has a unique solution defined in **R**. Since the function $v(4x_0 - x)$, $x \in (2x_0, 3x_0)$, is a solution of (2.34), this provides $v(x) = v(4x_0 - x)$, $\forall x \in (2x_0, 3x_0)$.

Now, we can repeat this procedure in the intervals $[nx_0, (n+1)x_0]$, $\forall n \in \mathbf{N}$, obtaining:

$$v(x) = -v(2x_0 - x), \ \forall \, x \in [x_0, 2x_0],$$

$$v(x) = v(4x_0 - x), \ \forall \, x \in [2x_0, 3x_0],$$

$$v(x) = -v(6x_0 - x), \ \forall \, x \in [3x_0, 4x_0],$$

$$\dots$$
(2.35)

The conclusion is that if v is a nontrivial solution of (2.26) for some $B \in \mathbf{R}^+$, and x_0 is the first zero point of v in $(0, L)$, then $L = 2nx_0$ for some $n \in \mathbf{N}$. Next we calculate $J_p(v)$.

It follows from previous reasonings that

$$J_p(v) = \frac{\int_0^L v'^2}{\left(\int_0^L |v|^{\frac{2p}{p-1}}\right)^{\frac{p-1}{p}}} = \frac{2n \int_0^{x_0} v'^2}{\left(2n \int_0^{x_0} |v|^{\frac{2p}{p-1}}\right)^{\frac{p-1}{p}}}.$$
(2.36)

From (2.28) we obtain

$$\int_0^{x_0} (v'(x))^2 \, dx = \frac{B(p-1)}{p} \left[-\int_0^{x_0} |v(x)|^{\frac{2p}{p-1}} \, dx + x_0 |v(0)|^{\frac{2p}{p-1}} \right]$$
(2.37)

and from (2.32) we obtain

$$\int_0^{x_0} |v(x)|^{\frac{2p}{p-1}} = \int_0^{x_0} \left(\frac{I}{x_0} \left(\frac{p}{B(p-1)} \right)^{1/2} \right)^{2p} \left[\varphi^{-1} \left(I - \frac{I}{x_0} x \right) \right]^{\frac{2p}{p-1}} dx.$$
(2.38)

Doing the change of variables $s = \varphi^{-1}(I(1 - \frac{x}{x_0}))$, we have

$$\int_0^{x_0} |v(x)|^{\frac{2p}{p-1}} = \left(\frac{I}{x_0} \left(\frac{p}{B(p-1)} \right)^{1/2} \right)^{2p} \frac{x_0}{I} \int_0^1 s^{\frac{2p}{p-1}} \left(1 - s^{\frac{2p}{p-1}} \right)^{-1/2} ds.$$
(2.39)

Integrating by parts the previous expression with $f(s) = s$, $g'(s) = s^{\frac{p+1}{p-1}}\left(1 - s^{\frac{2p}{p-1}}\right)^{-1/2}$, we deduce

$$\int_0^{x_0} |v(x)|^{\frac{2p}{p-1}} = \left(\frac{I}{x_0}\left(\frac{p}{B(p-1)}\right)^{1/2}\right)^{2p} \frac{x_0}{I}\frac{p-1}{2p-1}I. \tag{2.40}$$

If we substitute this expression in (2.37) and, moreover, we take into account (2.31), we obtain (think that $L = 2nx_0$)

$$\int_0^{x_0} |v'(x)|^2\, dx = \frac{B(p-1)}{p}x_0\left(\frac{I}{x_0}\left(\frac{p}{B(p-1)}\right)^{1/2}\right)^{2p}\frac{p}{2p-1}. \tag{2.41}$$

Now we can substitute (2.40) and (2.41) in (2.36). After some elementary calculations we deduce

$$J_p(v) = \frac{4n^2 I^2 p}{L^{2-\frac{1}{p}}(p-1)^{1-\frac{1}{p}}(2p-1)^{1/p}}. \tag{2.42}$$

At this point, one may observe two things. First, $J_p(v)$ does not depend on B. Second, all values of $n \in \mathbf{N}$ are possible in (2.42). In fact if $x_0 = \frac{L}{2n}$, formula (2.32) defines a nontrivial solution of (2.26). Therefore, the infimum m_p is attained if $n = 1$. Finally, doing the change of variables $s^{\frac{p}{p-1}} = \sin t$, we obtain $I = \int_0^1 \frac{ds}{\left(1 - s^{\frac{2p}{p-1}}\right)^{1/2}} = \frac{p-1}{p}K$, where $K = \int_0^{\pi/2}(\sin t)^{-1/p}\, dt$. This gives

$$m_p = \frac{4(p-1)^{1+\frac{1}{p}}}{L^{2-\frac{1}{p}}p(2p-1)^{1/p}}\left(\int_0^{\pi/2}(\sin x)^{-1/p}\, dx\right)^2. \tag{2.43}$$

Remark 2.8. In order to study other boundary conditions (Sect. 2.3), it seems essential to highlight the basic facts of the previous procedure.

We emphasize that if v is a nontrivial solution of (2.26) and x_0 is the first zero point of v in $(0, L)$, then $L = 2nx_0$ for some natural number $n \geq 1$ and, in addition,

$$v'(0) = v'(2x_0) = \ldots = v'(2nx_0) = 0,$$

$$v(x_0) = \ldots = v((2n-1)x_0) = 0, \tag{2.44}$$

and $v(x) \neq 0$, $v'(x) \neq 0$, $\forall\, x \in (jx_0, (j+1)x_0)$, $0 \leq j \leq 2n-1$. These properties allow to calculate, in a explicit way, the functions v' and v in $[0, L]$ and consequently, to find the value of $J_p(v)$ given in (2.42).

Remark 2.9. It is proved in [3] that β_p, as a function of $p \in [1, +\infty]$, is continuous.

Remark 2.10. If $L = 1$ and $1 \leq p < q < \infty$, then $\beta_p < \beta_q$ (see [3]). As a trivial consequence, if L is an arbitrary positive number, the mapping $(1, \infty) \to \mathbf{R}$, $p \to L^{-1/p}\beta_p$ is strictly increasing.

Now, we return to the linear boundary value problem (2.17), corollary 2.1, and remark 2.7. The following result establishes a *natural link* between the cases $p = 1$ and $p = \infty$. Previously, remember that from Theorems 2.1, 2.2 and 2.3 the constant β_p, defined as an infimum in 2.5, is attained, if and only if $1 < p \leq \infty$.

Corollary 2.2. *Let $a \in L^\infty \setminus \{0\}$, $0 \leq \int_0^L a(x)$, satisfying one of the following conditions:*

1. *$\|a^+\|_1 \leq \beta_1$,*
2. *There is some $p \in (1, \infty)$ such that $\|a^+\|_p < \beta_p$,*
3. *$\|a^+\|_\infty < \beta_\infty$ or $\|a^+\|_\infty = \beta_\infty$ and $a^+ \neq a_\infty$.*

Then for each $f \in L^\infty(0, L)$, the boundary value problem (2.17) has a unique solution.

Remark 2.11. We have shown that the best Sobolev constant β_p, defined in (2.5), can be computed by using a certain minimization problem given in Theorems 2.1, 2.2, and 2.3. Motivated by a completely different problem (an isoperimetric inequality known as Wulff theorem, of interest in crystallography), the authors studied in [9] a similar variational problem for the case of periodic or Dirichlet boundary conditions (see also [11] for more general minimization problems). Our treatment of the Euler equation associated with the mentioned minimization problem is different from that of Croce and Dacorogna [9].

2.2 Nonlinear Neumann Problems

Lyapunov inequalities can be used in the study of nonlinear resonant problems. To accomplish this, the linear results are combined with Schauder fixed point theorem.

We focus on a resonant nonlinear problem with Neumann boundary conditions, but the same ideas and methods can be used for other situations (see Sect. 2.3).

More precisely, let us consider the problem

$$u''(x) + f(x, u(x)) = 0, \ x \in (0, L), \ u'(0) = u'(L) = 0, \tag{2.45}$$

where $f : [0, L] \times \mathbf{R} \to \mathbf{R}$, $(x, u) \to f(x, u)$ is continuous.

The *associated linear problem*

$$u''(x) = 0, \ x \in (0, L), \ u'(0) = u'(L) = 0 \tag{2.46}$$

has nontrivial solutions (any constant function) and this is the reason why we call (2.45) a resonant problem.

If (2.45) is linear, i.e., it is of the type

$$u''(x) + a(x)u(x) = 0, \ x \in (0, L), \ u'(0) = u'(L) = 0 \tag{2.47}$$

and for some integer $n \geq 0$ there is a positive number δ such that

$$\lambda_n + \delta \leq a(x) \leq \lambda_{n+1} - \delta, \ \text{in} \ [0, L], \tag{2.48}$$

where λ_n is an eigenvalue of the eigenvalue problem (2.2), then (2.47) has only the trivial solution $u \equiv 0$ (see, for instance, [20]). In particular, for the first eigenvalue $\lambda_0 = 0$, (2.48) becomes

$$\delta \leq a(x) \leq \frac{\pi^2}{L^2} - \delta, \ \text{in} \ [0, L]. \tag{2.49}$$

We must remark that (2.48) does not allow to the function $a(\cdot)$ to cross any eigenvalue of (2.2). Using Lyapunov inequalities, it is possible that $f_u(x, u)$ in (2.45) crosses the eigenvalues λ_n (f_u means the partial derivative of the function $f(x, u)$ with respect to the variable u.) and it is possible to provide some extensions of Corollary 2.2 to nonlinear situations.

To this respect, we will assume throughout this section that the following hypothesis is satisfied

(H) f, f_u are continuous on $[0, L] \times \mathbf{R}$ and $0 \leq f_u(x, u)$ on $[0, L] \times \mathbf{R}$.

Then, the existence of a solution u of (2.45) implies

$$\int_0^L f(x, u(x)) \, dx = 0. \tag{2.50}$$

Now, the previous hypothesis **(H)** implies that $f(x, u)$ is increasing with respect to u. Therefore,

$$\int_0^L f(x, m) \, dx \leq \int_0^L f(x, u(x)) \, dx = 0 \leq \int_0^L f(x, M) \, dx,$$

where $m = \min_{[0,L]} u$ and $M = \max_{[0,L]} u$ and consequently

$$\int_0^L f(x, z) \, dx = 0 \tag{2.51}$$

for some $z \in \mathbf{R}$. However, conditions (H) and (2.51) are not sufficient for the existence of solutions of (2.45). Indeed, if $n \in \mathbf{N}$ is any natural number, consider the problem

$$u''(x) + n^2\pi^2 u(x) + \cos(n\pi x) = 0, \ x \in (0, 1), \ u'(0) = u'(1) = 0. \tag{2.52}$$

The function $f(x, u) = n^2\pi^2 u + \cos(n\pi x)$ satisfies (H) and (2.51), but the Fredholm alternative theorem [16] shows that there is no solution of (2.52).

If (H) and (2.51) are assumed, and for instance, $L = 1$ for simplicity, different supplementary assumptions can be given which imply the existence of a solution of (2.45). For example

(h1) $f_u(x, u) \leq \beta(x)$ on $[0, 1] \times \mathbf{R}$ with $\beta \in L^\infty(0, 1)$, $\beta(x) \leq \pi^2$ on $[0, 1]$ and $\beta(x) < \pi^2$ on a subset of $(0, 1)$ of positive measure.

Conditions of this type are referred to as nonuniform nonresonance conditions with respect to the first positive eigenvalue of the associated linear homogeneous problem. By using variational methods, it is proved in [26] that (H), (2.51), and (h1) imply the existence of solutions of (2.45). Restriction (h1) is related to Lyapunov-type inequalities: the number π^2 is the best L_∞ Lyapunov constant, β_∞, for $L = 1$ (Theorem 2.1).

On the other hand, in [18] it is supposed

(h2) $f_u(x, u) \leq \beta(x)$ on $[0, 1] \times \mathbf{R}$ with $\beta \in L^1(0, 1)$ and $\displaystyle\int_0^1 \beta(x)\, dx \leq 4$

The authors use Optimal Control theory methods to prove that (H), (2.51), and (h2) imply the existence and uniqueness of solutions of (2.45). Restriction (h2) is also related to Lyapunov-type inequalities: the number 4 is the best L_1 Lyapunov constant, β_1, for $L = 1$ (Theorem 2.2).

Let us observe that supplementary conditions (h1) and (h2) are given respectively in terms of $\|\beta\|_\infty$ and $\|\beta\|_1$, the usual norms in the spaces $L^\infty(0, 1)$ and $L^1(0, 1)$. Also, it is trivial that under the hypotheses (H) and (2.51), (h1) and (h2) are not related (i.e., none of these hypotheses implies the other).

In the next theorem we provide supplementary conditions in terms of $\|\beta\|_p$, $1 \leq p \leq \infty$. As a consequence, a natural relation between (h1) and (h2) arises if one takes into account Remark (2.9) and studies the limits of $\|\beta\|_p$ for $p \to 1^+$ and $p \to \infty$ (see [3] for further details).

Theorem 2.4. *Let us consider (2.45) where the following requirements are fulfilled:*

1. f and f_u are continuous on $[0, L] \times \mathbf{R}$.
2. $0 \leq f_u(x, u)$ in $[0, L] \times \mathbf{R}$. Moreover, for each $u \in C[0, L]$ one has $f_u(x, u(x)) \neq 0$, a.e. on $[0, L]$ and $\displaystyle\int_0^L f(x, 0)\, dx = 0$.
3. For some function $\beta \in L^\infty(0, L)$, we have $f_u(x, u) \leq \beta(x)$ on $[0, L] \times \mathbf{R}$ and β satisfies some of the conditions given in Corollary 2.2.

Then, problem (2.45) has a unique solution.

Proof. The proof consists of two parts: existence and uniqueness of solutions of (2.45). We begin with the second one.

Uniqueness of Solutions We assume that (2.45) has two solutions. Then, the mean value theorem [8] and Corollary (2.2) are used to prove that they are the same.

Let u_1 and u_2 be two solutions of (2.45). Then,

$$- (u_1 - u_2)''(x) = f(x, u_1(x)) - f(x, u_2(x))$$

$$= \int_0^1 \frac{d}{d\theta} [f(x, u_2(x) + \theta(u_1(x) - u_2(x)))] \, d\theta$$

$$= \left[\int_0^1 f_u(x, u_2(x) + \theta(u_1(x) - u_2(x))) \, d\theta \right] (u_1(x) - u_2(x)), \ x \in [0, L].$$

$$\tag{2.53}$$

Hence, the function $u = u_1 - u_2$ is a solution of a homogeneous problem of the type (2.17) with $a(x) = \int_0^1 f_u(x, u_2(x) + \theta u(x)) \, d\theta$. From the hypotheses of the theorem and applying Corollary 2.2, we obtain $u \equiv 0$.

Existence of Solutions The main idea is to rewrite (2.45) in an equivalent form, such that the solutions of (2.45) be the fixed points of a certain completely continuous operator, and then, to apply the Schauder fixed point theorem [12]. To see this, by using the same idea that in (2.53), we rewrite (2.45) as

$$0 = u''(x) + f(x, u(x)) = u''(x) + f(x, u(x)) - f(x, 0) + f(x, 0)$$

$$= u''(x) + \int_0^1 \frac{d}{d\theta} [f(x, \theta u(x))] \, d\theta + f(x, 0)$$

$$= u''(x) + \left[\int_0^1 f_u(x, \theta u(x)) \, d\theta \right] u(x) + f(x, 0). \tag{2.54}$$

Therefore, u is a solution of (2.45) if and only if u satisfies

$$u''(x) + b(x, u(x))u(x) = -f(x, 0), \ x \in [0, L], \ u'(0) = u'(L) = 0, \tag{2.55}$$

where the continuous function $b : [0, L] \times \mathbf{R} \to \mathbf{R}$ is defined by

$$b(x, z) = \int_0^1 f_u(x, \theta z) \, d\theta.$$

From the hypotheses of the theorem, it is deduced that for each function $y \in C^1([0, L], \mathbf{R})$, the linear equation

$$u''(x) + b(x, y(x))u(x) = -f(x, 0), \ x \in [0, L], \ u'(0) = u'(L) = 0 \tag{2.56}$$

satisfies all the hypotheses of Corollary 2.2 and consequently, (2.56) has a unique solution u_y. Then, if $X = C^1([0, L], \mathbf{R})$ with the usual norm, i.e.,

$$\|y\|_X = \max_{x \in [0,L]} |y(x)| + \max_{x \in [0,L]} |y'(x)|, \ \forall \, y \in X$$

we can define the operator $T : X \to X$, by $Ty = u_y$. Clearly, u is a solution of (2.45) if and only if y is a fixed point of T.

We claim that T is completely continuous (T is continuous and if $B \subset X$ is bounded, then $T(B)$ is relatively compact in X) and that $T(X)$ is bounded. Then, the Schauder fixed point theorem ensures that T has a fixed point which provides a solution of (2.45).

To prove the claim, if $T(X)$ is not bounded, there would exist a sequence $\{y_n\} \subset X$ such that $\|u_{y_n}\|_X \to \infty$. Moreover, from the hypotheses of the theorem, the sequence of functions $\{b(\cdot, y_n(\cdot))\}$ is bounded in $L^2(0, L)$ and, passing to a subsequence if necessary, we may assume that $\{b(\cdot, y_n(\cdot))\}$ is weakly convergent in $L^2(0, L)$ to a function β_0 satisfying $0 \le \beta_0(x) \le \beta(x)$, a.e. in $[0, L]$ (see [2] for the properties of the convergence in $L^2(0, L)$).

In addition, each u_{y_n} satisfies

$$u_{y_n}''(x) + b(x, y_n(x))u_{y_n}(x) = -f(x, 0), \ x \in [0, L], \ u'(0) = u'(L) = 0. \qquad (2.57)$$

Since the embedding $H^1(0, L) \subset C[0, L]$ is compact (in $C[0, L]$ we take the uniform norm), if $z_n \equiv \dfrac{u_{y_n}}{\|u_{y_n}\|_X}$, then passing to a subsequence if necessary, we may assume that $z_n \to z_0$, uniformly in $[0, L]$, where z_0 satisfies $\|z_0\|_X = 1$ and

$$z_0''(x) + \beta_0(x)z_0(x) = 0, \ x \in [0, L], \ z_0'(0) = z_0'(L) = 0. \qquad (2.58)$$

Moreover, from the hypotheses of the theorem, we have for each $n \in \mathbf{N}$,

$$\int_0^L b(x, y_n(x))u_{y_n}(x)\, dx = - \int_0^L f(x, 0)\, dx = 0.$$

Also, the function $b(\cdot, y_n(\cdot))$ is nonnegative and not identically zero. Therefore, for each $n \in \mathbf{N}$, the function u_{y_n} has a zero in $[0, L]$. This implies that for each $n \in \mathbf{N}$, the function z_n has a zero in $[0, L]$ and hence so does z_0. Taking into account (2.58), $\beta_0 \in L^\infty(0, L) \setminus \{0\}$. This is a contradiction with Corollary 2.2.

Now, let us prove that the operator T is continuous. To see this, if $\{y_n\} \to y_0$ in the space X and u_{y_n} does not converge to u_{y_0}, passing to a subsequence if necessary, there exists a constant $\delta > 0$ such that $u_{y_n} \notin B_X(u_{y_0}; \delta)$, $\forall n \in \mathbf{N}$, where $B_X(u_{y_0}; \delta)$ denotes the open ball in X of center u_{y_0} and radius δ. Also, taking into account (2.56) and the boundness of the operator T, we obtain that the sequence $\{u_{y_n}''\}$ is uniformly bounded. Thus, by Arzela–Ascoli Theorem [7], again passing to a subsequence if necessary, we deduce that u_{y_n} converges to some function u_0. But, by the uniqueness of solution for problem (2.56), we must have $u_0 = u_{y_0}$, which is a contradiction.

Finally, by using again the Arzela–Ascoli theorem, it is trivial from (2.56) that if $B \subset X$ is bounded, then $T(B)$ is relatively compact in X.

Remark 2.12. If $f(x, u) = a(x)u$, the second hypothesis in the previous theorem becomes $0 \le a(x)$ and $a(x) \neq 0$, a.e. on $[0, L]$.

Remark 2.13. Since the change of variables $u(x) = v(x) + z$, $z \in \mathbf{R}$, transforms (2.45) into the problem

$$v''(x) + f(x, v(x) + z) = 0, \ x \in (0, L), \ v'(0) = v'(L) = 0,$$

the condition $\int_0^L f(x, 0)\, dx = 0$ in the previous theorem may be substituted by

$\int_0^L f(x, z)\, dx = 0$, for some $z \in \mathbf{R}$.

Remark 2.14. Taking into account Remark 2.9, previous result establishes a clear relationship between Theorem B in [18] and Theorem 2 in [26] for the case of ordinary differential equations.

Remark 2.15. Let us remark that the hypothesis of the previous theorem allows the function $f_u(x, u)$ to cross an arbitrary number of different eigenvalues λ_n of the eigenvalue problem (2.2) (see [3, 18]).

2.3   The Variational Method for Other Boundary Conditions

The variational method that we have used in Sect. 2.1 (Theorem 2.3), to obtain the explicit value of the constant β_p, $1 < p < \infty$, is valid for many other boundary conditions. Remember the two key points for Neumann problem (2.1).

1. The set of boundary value problems

$$v''(x) + B|v(x)|^{\frac{2}{p-1}} v(x) = 0, \ x \in (0, L), \ v'(0) = v'(L) = 0, \ B \in \mathbf{R}^+ \tag{2.59}$$

provides

$$\beta_p = \inf_{B \in \mathbf{R}^+} \ \inf_{v \in S_B} \ J_p(v), \tag{2.60}$$

where

$$J_p(v) = \frac{\displaystyle\int_0^L v'^2}{\left(\displaystyle\int_0^L |v|^{\frac{2p}{p-1}}\right)^{\frac{p-1}{p}}}$$

and for a given $B \in \mathbf{R}^+$, S_B denotes the set of all nontrivial solutions of (2.59).

2. If v is a nontrivial solution of (2.59) for some $B \in \mathbf{R}^+$, then

$$J_p(v) = \frac{4n^2 I^2 p}{L^{2-\frac{1}{p}}(p-1)^{1-\frac{1}{p}}(2p-1)^{1/p}}, \tag{2.61}$$

where

$$I = \frac{p-1}{p} \int_0^{\pi/2} (\sin x)^{-1/p}\, dx \tag{2.62}$$

and n is the unique natural number (depending on v), satisfying the properties:

$$
\begin{aligned}
&x_0 \text{ is the first zero point of } v \text{ in } (0, L), \ L = 2nx_0, \\
&v'(0) = v'(2x_0) = \ldots = v'(2nx_0) = 0, \\
&v(x_0) = \ldots = v((2n-1)x_0) = 0, \\
&v(x) \neq 0, \ v'(x) \neq 0, \ \forall \, x \in (jx_0, (j+1)x_0), \ 0 \leq j \leq 2n-1.
\end{aligned}
\tag{2.63}
$$

Let us emphasize that the value of $J_p(v)$ in (2.61) does not depend, explicitly, on the positive constant B and that to obtain β_p we must find the minimum value of n in the expression (2.61). For instance, for Neumann boundary conditions this minimum value is $n = 1$ (see the last part of Theorem 2.3).

Below we describe the main ideas for other boundary conditions.

In the remainder of the chapter we will denote as β_p^N the constant β_p obtained above for Neumann boundary conditions.

Dirichlet Boundary Conditions This case is very similar to the Neumann one. If we consider the linear problem

$$
u''(x) + a(x)u(x) = 0, \ x \in (0, L), \ u(0) = u(L) = 0, \tag{2.64}
$$

where $a \in \Lambda^D$ and Λ^D is defined by

$$
\Lambda^D = \{a \in L^1(0, L) \text{ such that (2.64) has nontrivial solutions }\} \tag{2.65}
$$

then, for each p with $1 \leq p \leq \infty$, we can define the functional $I_p : \Lambda^D \cap L^p(0, L) \to \mathbf{R}$ given by $I_p(a) = \|a^+\|_p$ (the same expression as in (2.4)), and in a similar form, we can define the constant

$$
\beta_p^D \equiv \inf_{a \in \Lambda^D \cap L^p(0,L)} I_p(a), \ 1 \leq p \leq \infty. \tag{2.66}
$$

Taking into account the same ideas that for the Neumann problem, it can be easily proved that

$$
\beta_p^D = \beta_p^N, \ 1 \leq p \leq \infty. \tag{2.67}
$$

In the proof, we must simply replace the spaces X_p of Theorems 2.1–2.3 by the Sobolev space $H_0^1(0, L)$ and (2.63) by

$$
\begin{aligned}
&v(0) = v(2x_0) = \ldots = v(2nx_0) = 0, \\
&v'(x_0) = \ldots = v'((2n-1)x_0) = 0, \\
&v(x) \neq 0, \ v'(x) \neq 0, \ \forall \, x \in (jx_0, (j+1)x_0), \ 0 \leq j \leq 2n-1.
\end{aligned}
\tag{2.68}
$$

Remark 2.16. Let us note that, contrary to what happens for Neumann problems, in the minimization problems associated with Dirichlet boundary conditions, we do not need to impose any additional restriction to the space $H_0^1(0, L)$ (see [28]). This is due to the fact that the homogeneous linear part of (2.64)

$$u''(x) = 0, \ x \in (0, L), \ u(0) = u(L) = 0 \tag{2.69}$$

has only the trivial solution $u \equiv 0$. In this work, we will call to this type of problems *nonresonant problems*.

Periodic Boundary Conditions In the case of the periodic boundary value problem

$$u''(t) + a(t)u(t) = 0, \ t \in (0, T), \ u(0) - u(T) = u'(0) - u'(T) = 0 \tag{2.70}$$

we assume that $a \in L_T(\mathbf{R}, \mathbf{R})$, the set of T-periodic functions $a : \mathbf{R} \to \mathbf{R}$ such that $a|_{[0,T]} \in L^1(0, T)$ (due to the applications to stability, it is convenient to use t as the independent variable, instead of x).

If we define the set

$$\Lambda^{\mathrm{per}} = \{a \in L_T(\mathbf{R}, \mathbf{R}) \setminus \{0\} : \int_0^T a(t) \, dt \geq 0 \text{ and (2.70) has nontrivial solutions } \}$$
$$\tag{2.71}$$

the positive eigenvalues of the eigenvalue problem

$$u''(t) + \lambda u(t) = 0, \ t \in (0, T), \ u(0) - u(T) = u'(0) - u'(T) = 0 \tag{2.72}$$

belong to Λ^{per}. Therefore, for each p with $1 \leq p \leq \infty$, we can define the L^p Lyapunov constant for the periodic problem, β_p^{per}, as the real number

$$\beta_p^{\mathrm{per}} \equiv \inf_{a \in \Lambda^{\mathrm{per}} \cap L^p(0,T)} \|a^+\|_p. \tag{2.73}$$

An explicit expression for the constant β_p^{per}, as a function of p and T, has been obtained in [30]. As in the Neumann case, we can obtain a characterization of β_p^{per} as a minimum of a convenient minimization problem, where only some appropriate subsets of the space $H^1(0, T)$ are used (see [6] for further details).

Since (2.72) is, as (2.1), a resonant problem, just to get a variational characterization of β_p^{per} we need an additional restriction to the space $H^1(0, T)$. This is shown in the next theorem.

Theorem 2.5. *If $1 \leq p \leq \infty$ is a given number, let us define the sets X_p^{per} and the functionals $I_p^{\mathrm{per}} : X_p^{\mathrm{per}} \setminus \{0\} \to \mathbf{R}$ as*

$$X_1^{\mathrm{per}} = \{v \in H^1(0, T) : v(0) - v(T) = 0, \ \max_{t \in [0,T]} v(t) + \min_{t \in [0,T]} v(t) = 0\},$$

$$X_p^{\mathrm{per}} = \left\{v \in H^1(0, T) : v(0) - v(T) = 0, \int_0^T |v|^{\frac{2}{p-1}} v = 0\right\}, \ \textit{if } 1 < p < \infty,$$

$$X_\infty^{per} = \{v \in H^1(0,T) : v(0) - v(T) = 0, \int_0^T v = 0\},$$

$$I_1^{per}(v) = \frac{\int_0^T v'^2}{\|v\|_\infty^2}, \quad I_p^{per}(v) = \frac{\int_0^T v'^2}{\left(\int_0^T |v|^{\frac{2p}{p-1}}\right)^{\frac{p-1}{p}}}, \; if \; 1 < p < \infty, \quad I_\infty^{per}(v) = \frac{\int_0^T v'^2}{\int_0^T v^2}.$$

(2.74)

Then, the L_p Lyapunov constant β_p^{per} defined in (2.73), satisfies

$$\beta_p^{per} = \min_{X_p^{per}\setminus\{0\}} I_p^{per}, \; 1 \le p \le \infty.$$

(2.75)

Proof. Only those innovative details with respect to the Neumann case are shown [6].

The case $p = 1$. It is very well known that $\beta_1^{per} = \frac{16}{T}$ [17, 30]. Now, if $u \in X_1^{per} \setminus \{0\}$, then there exists $x_0 \in [0, T]$ such that $u(x_0) = 0$. Taking into account that u can be extended as a T-periodic function to **R**, if we define the function $v(x) = u(x + x_0), \; \forall \, x \in \mathbf{R}$, then $v|_{[0,T]} \in X_1^{per} \setminus \{0\}$, $v(0) = v(T) = 0$ and $I_1^{per}(u) = I_1^{per}(v)$. In addition (if it is necessary, we can choose $-v$ instead of v), there exist $0 < x_1 < x_2 < x_3 < T$ such that

$$v(x_1) = \max_{[0,T]} v, \quad v(x_2) = 0, \quad v(x_3) = \min_{[0,T]} v.$$

If $x_0 = 0, x_4 = T$, it follows from the Cauchy–Schwarz inequality

$$\int_0^T v'^2 = \sum_{i=0}^3 \int_{x_i}^{x_{i+1}} v'^2 \ge \sum_{i=0}^3 \frac{\left(\int_{x_i}^{x_{i+1}} |v'|\right)^2}{x_{i+1} - x_i}$$

$$\ge \sum_{i=0}^3 \frac{\left(\int_{x_i}^{x_{i+1}} v'\right)^2}{x_{i+1} - x_i} = \sum_{i=0}^3 \frac{(v(x_{i+1}) - v(x_i))^2}{x_{i+1} - x_i}$$

$$= \|v\|_\infty^2 \sum_{i=0}^3 \frac{1}{x_{i+1} - x_i} \ge \frac{16}{T} \|v\|_\infty^2.$$

(2.76)

Consequently

$$I_1^{per}(u) = \frac{\int_0^T u'^2}{\|u\|_\infty^2} = I_1^{per}(v) \ge \frac{16}{T}, \; \forall \, u \in X_1^{per} \setminus \{0\}.$$

(2.77)

On the other hand, the function $w \in X_1^{per} \setminus \{0\}$ defined as

$$w(x) = \begin{cases} x, & \text{if } 0 \le x \le T/4, \\ -(x - \frac{T}{2}), & \text{if } T/4 \le x \le 3T/4, \\ (x - T), & \text{if } 3T/4 \le x \le T, \end{cases} \tag{2.78}$$

satisfies

$$\frac{\int_0^T w'^2}{\|w\|_\infty^2} = \frac{16}{T}.$$

Consequently, the case $p = 1$ is proved.

The case $p = \infty$. It is very well known that $\beta_\infty^{per} = \frac{4\pi^2}{T^2}$, the first positive eigenvalue of the eigenvalue problem (2.72) (see [30]). From its variational characterization, we obtain

$$\beta_\infty^{per} = \min_{X_\infty^{per} \setminus \{0\}} I_\infty^{per}.$$

The case $1 < p < \infty$. The ideas are similar to those used in the case of Neumann boundary conditions. If we denote

$$m_p^{per} = \inf_{X_p^{per} \setminus \{0\}} I_p^{per}$$

then this infimum is attained in some function u_0 which satisfies

$$u_0''(x) + A_p(u_0)|u_0(x)|^{\frac{2}{p-1}} u_0(x) = 0, \; x \in (0, T), \tag{2.79}$$
$$u_0(0) - u_0(T) = 0, \; u_0'(0) - u_0'(T) = 0,$$

where

$$A_p(u_0) = m_p^{per} \left(\int_0^T |u_0|^{\frac{2p}{p-1}} \right)^{\frac{-1}{p}}. \tag{2.80}$$

Let us observe that the previous equation is of the type (2.59), but with periodic boundary conditions instead of Neumann ones. As it was commented at the beginning of this section, this is not a problem. If one has an exact knowledge about the number and distribution of the zeros of the functions u_0 and u_0', the Euler equation (2.79) can be integrated (see [3], Lemma 2.7). In our case, it is not restrictive to assume $u_0(0) = u_0(T) = 0$ (see the previous case $p = 1$). Then, if we denote the zeros of u_0 in $[0, T]$ by $0 = x_0 < x_2 < \ldots < x_{2n} = T$ and the zeros of u_0' in $(0, T)$ by $x_1 < x_3 < \ldots < x_{2n-1}$, we obtain

$$m_p^{per} = \frac{4n^2 l^2 p}{T^{2 - \frac{1}{p}} (p - 1)^{1 - \frac{1}{p}} (2p - 1)^{1/p}}, \tag{2.81}$$

where I is defined in (2.62).

The novelty here is that, for the periodic boundary value problem (2.79), $n \geq 2$ (see the relations (2.35)), while for the Neumann and Dirichlet problem $n \geq 1$.

The conclusion is that

$$m_p^{\text{per}} = \frac{16l^2 p}{T^{2-\frac{1}{p}}(p-1)^{1-\frac{1}{p}}(2p-1)^{1/p}} \tag{2.82}$$

that is, four times the corresponding L^p Lyapunov constant for the Dirichlet and the Neumann problem. Finally, in [30] it is shown that this is, exactly, the L^p Lyapunov constant for the periodic problem. Consequently, $m_p^{\text{per}} = \beta_p^{\text{per}}$, $1 < p < \infty$.

Finally, we treat in this section with antiperiodic boundary conditions, another important case due to its applications to stability theory (Chap. 3). As we will show, in some aspects this case is similar to the case of periodic boundary conditions, but in others it is similar to Neumann or Dirichlet boundary conditions.

Antiperiodic Boundary Conditions   Let us consider the antiperiodic boundary value problem

$$u''(t) + a(t)u(t) = 0, \; t \in (0, T), \; u(0) + u(T) = u'(0) + u'(T) = 0 \tag{2.83}$$

where $a \in L_T(\mathbf{R}, \mathbf{R})$.

If we define the set

$$\Lambda^{\text{ant}} = \{a \in L_T(\mathbf{R}, \mathbf{R}) : (2.83) \text{ has nontrivial solutions }\} \tag{2.84}$$

the positive eigenvalues of the eigenvalue problem

$$u''(t) + \lambda u(t) = 0, \; t \in (0, T), \; u(0) + u(T) = u'(0) + u'(T) = 0 \tag{2.85}$$

belong to Λ^{ant}. Therefore, for each p with $1 \leq p \leq \infty$, we can define the L^p Lyapunov constant for the antiperiodic problem, β_p^{ant}, as the real number

$$\beta_p^{\text{ant}} \equiv \inf_{a \in \Lambda^{\text{ant}} \cap L^p(0,T)} \|a^+\|_p \tag{2.86}$$

An explicit expression for the constant β_p^{ant}, as a function of p and T, has been obtained in [30]. As in the cases of Neumann, Dirichlet, or periodic boundary conditions, it is possible to prove a characterization of β_p^{ant} as a minimum of a convenient minimization problem, where only some appropriate subsets of the space $H^1(0, T)$ are used (see [6] for further details). Since (2.83) is, as (2.64), a no resonant problem, i.e., the linear part

$$u''(t) = 0, \; t \in (0, T), \; u(0) + u(T) = u'(0) + u'(T) = 0 \tag{2.87}$$

has only the trivial solution, just to get a variational characterization of β_p^{ant} we do not need any additional restriction to the space $H^1(0, T)$, except $u(0) + u(T) = 0$. This is shown in the next theorem, where the proof is omitted (see [6]).

Theorem 2.6. *If* $1 \le p \le \infty$ *is a given number, let us define the sets* X_p^{ant} *and the functional* $I_p^{ant} : X_p^{ant} \setminus \{0\} \to \mathbf{R}$, *as*

$$X_p^{ant} = \left\{ v \in H^1(0,T) : v(0) + v(T) = 0 \right\}, \ 1 \le p \le \infty,$$

$$I_1^{ant}(v) = \frac{\displaystyle\int_0^T v'^2}{\|v\|_\infty^2}, \ I_p^{ant}(v) = \frac{\displaystyle\int_0^T v'^2}{\left(\displaystyle\int_0^T |v|^{\frac{2p}{p-1}}\right)^{\frac{p-1}{p}}}, \ \textit{if} \ 1 < p < \infty, \ I_\infty^{ant}(v) = \frac{\displaystyle\int_0^T v'^2}{\displaystyle\int_0^T v^2}.$$

$$\text{(2.88)}$$

Then, the L_p *Lyapunov constant* β_p^{ant} *defined in* (2.86) *satisfies*

$$\beta_p^{ant} = \min_{X_p^{ant}\setminus\{0\}} I_p^{ant}, \ 1 \le p \le \infty. \tag{2.89}$$

Remark 2.17. Using the procedure described in Sect. 2.1 of this chapter, many other boundary conditions can be studied. We bring out the case of problems of mixed type

$$u''(x) + a(x)u(x) = 0, \ x \in (0,L), \ u'(0) = u(L) = 0,$$

where the number n of the relation (2.61) must be chosen as $n = 1/2$. However, due to the important relationship of this case with the notion of disfocality and its applications to resonant nonlinear problems and the theory of stability, such problems will be treated in the next section.

2.4 Disfocality

Under the natural restrictions $a \in L^1(0,L) \setminus \{0\}$ and $\displaystyle\int_0^L a(x) \, dx \ge 0$, the relation between Neumann boundary conditions and disfocality arises in a natural way, since if $u \in H^1(0,L)$ is any nontrivial solution of

$$u''(x) + a(x)u(x) = 0, \ x \in (0,L), \ u'(0) = u'(L) = 0 \tag{2.90}$$

then u must have a zero c in the interval $(0,L)$. In fact, $u(0) \ne 0$ and $u(L) \ne 0$. Then, if u has not zeros in the interval $(0,L)$, we can assume that u is, for example, a positive (nonconstant) solution of (2.90). Considering $v = \dfrac{1}{u}$ as test function in the weak formulation of (2.90), we obtain

$$\int_0^L a = \int_0^L au\frac{1}{u} = \int_0^L u'(\frac{1}{u})' = -\int_0^L \frac{u'^2}{u^2} < 0$$

which is a contradiction with the hypothesis $\int_0^L a(x)\,dx \geq 0$.

In consequence both problems

$$v''(x) + a(x)v(x) = 0, \ x \in (0, c), \ v'(0) = v(c) = 0 \qquad\qquad \textbf{PM(0,c)}$$

and

$$v''(x) + a(x)v(x) = 0, \ x \in (c, L), \ v(c) = v'(L) = 0 \qquad\qquad \textbf{PM(c,L)}$$

have nontrivial solutions.

This simple observation (which has been previously employed in the case of Dirichlet boundary conditions, [15, 19]) can be used to deduce the following conclusion: if $a \in L^1(0, L) \setminus \{0\}$ with $\int_0^L a \geq 0$ is any function such that for any $c \in (0, L)$, either problem **PM(0,c)** or problem **PM(c,L)** has only the trivial solution, then problem (2.90) has only the trivial solution.

Below we study the relation between the best L_p Lyapunov constants for the problems (2.90) and

$$u''(x) + a(x)u(x) = 0, \ x \in (0, L), \ u'(0) = u(L) = 0, \qquad (2.91)$$

where for Neumann problem (2.90), function $a \in \Lambda$ and Λ is defined by

$$\Lambda = \{a \in L^1(0, L) \setminus \{0\} : \int_0^L a(x)\,dx \geq 0 \text{ and (2.90) has nontrivial solutions}\} \qquad (2.92)$$

whereas for mixed problem (2.91), function $a \in \Lambda^*$ and Λ^* is defined by

$$\Lambda^* = \{a \in L^1(0, L) : (2.91) \text{ has nontrivial solutions}\} \qquad (2.93)$$

Here $u \in H = H^1(0, L)$ (the usual Sobolev space) in the case of Neumann conditions and $u \in H^* = \{u \in H : u(L) = 0\}$ in the case of mixed boundary conditions. Obviously, the positive eigenvalues of the problems

$$u''(x) + \lambda u(x) = 0, \ x \in (0, L), \ u'(0) = u'(L) = 0 \qquad (2.94)$$

and

$$u''(x) + \lambda u(x) = 0, \ x \in (0, L), \ u'(0) = u(L) = 0 \qquad (2.95)$$

belong, respectively, to Λ and Λ^*. Therefore Λ and Λ^* are both not empty and the quantities

$$\beta_p \equiv \inf_{a \in \Lambda \cap L^p(0,L)} \|a^+\|_{L^p(0,L)}, \ 1 \leq p \leq \infty \tag{2.96}$$

and

$$\beta_p^* \equiv \inf_{a \in \Lambda^* \cap L^p(0,L)} \|a^+\|_{L^p(0,L)}, \ 1 \leq p \leq \infty \tag{2.97}$$

are well defined.

The next theorem establishes a clear relation between β_p and β_p^*. When it is necessary, we will write $\Lambda(0,L), \beta_p(0,L), \dots$ to show up the explicit dependence of these quantities with respect to the interval $(0,L)$. Also, it is possible to define, in an analogous manner, $\Lambda(c,d), \beta_p(c,d), \dots$ for arbitrary real numbers $c < d$.

Theorem 2.7. *If* $1 \leq p \leq \infty$, *we have* $\beta_p^* = \beta_p/4$.

Proof. By using the definition of β_p and β_p^*, and doing a trivial change of variables, it is easy to prove the equalities

$$\beta_p(0,c) = \left(\frac{c}{d}\right)^{\frac{1}{p}-2} \beta_p(0,d), \ \forall c, d \in \mathbf{R}^+, \ \forall p, \ 1 \leq p \leq \infty \tag{2.98}$$

and

$$\beta_p^*(0,c) = \left(\frac{c}{d}\right)^{\frac{1}{p}-2} \beta_p^*(0,d), \ \forall c, d \in \mathbf{R}^+, \ \forall p, \ 1 \leq p \leq \infty. \tag{2.99}$$

Moreover, problem (2.91) becomes

$$v''(x) + a(L-x)v(x) = 0, \ x \in (0,L), \ v(0) = v'(L) = 0 \tag{2.100}$$

through the variable change $y = L - x$ and it is clear that

$$\|a(\cdot)\|_{L^p(0,L)} = \|a(L - \cdot)\|_{L^p(0,L)}$$

Lemma 2.1. *If* $1 \leq p \leq \infty$, *we have* $\beta_p^* \leq \beta_p/4$.

Proof. If $a \in \Lambda(0,L) \cap L^p(0,L)$ and u is a nontrivial solution of (2.90), there exists $c \in (0,L)$ such that $u(c) = 0$. Therefore both problems **PM(0,c)** and **PM(c,L)** have nontrivial solutions. In consequence, function a, restricted to the interval $[0,c]$ belongs to $\Lambda^*(0,c)$ and function $a(L+c-\cdot)$, restricted to the interval $[c,L]$ belongs to $\Lambda^*(c,L)$. Let us assume $1 \leq p < \infty$. Then taking into account the definition of β_p^*, (2.99) and (2.100), we have

$$\|a^+\|_{L^p(0,L)}^p = \|a^+\|_{L^p(0,c)}^p + \|a^+\|_{L^p(c,L)}^p$$

$$\geq (\beta_p^*(0,c))^p + (\beta_p^*(c,L))^p$$

$$= \left(\frac{L}{c}\right)^{2p-1} (\beta_p^*(0,L))^p + \left(\frac{L}{L-c}\right)^{2p-1} (\beta_p^*(0,L))^p$$

$$= \left[\left(\frac{L}{c}\right)^{2p-1} + \left(\frac{L}{L-c}\right)^{2p-1}\right] (\beta_p^*(0,L))^p$$

$$\geq (\inf_{c\in(0,L)} g(c))(\beta_p^*(0,L))^p, \tag{2.101}$$

where $g : (0,L) \to \mathbf{R}$ is defined by

$$g(c) = \left[\left(\frac{L}{c}\right)^{2p-1} + \left(\frac{L}{L-c}\right)^{2p-1}\right], \ \forall c \in (0,L).$$

It is easily checked that

$$g'(c) < 0, \ \forall c \in (0, L/2) \text{ and } g'(c) > 0, \ \forall c \in (L/2, L).$$

Thus

$$\inf_{c\in(0,L)} g(c) = g(L/2) = 4^p. \tag{2.102}$$

Therefore, from (2.101) and (2.102) we deduce

$$\|a^+\|_{L^p(0,L)}^p \geq 4^p(\beta_p^*(0,L))^p, \ \forall a \in \Lambda \cap L^p(0,L).$$

Similar ideas may be used in the case $p = \infty$. This proves the lemma.

Lemma 2.2. *If* $1 \leq p \leq \infty$, *we have* $\beta_p^* \geq \beta_p/4$.

Proof. If $a \in \Lambda^*(0,L) \cap L^p(0,L)$ and u is a nontrivial solution of (2.91), let us define the functions

$$\tilde{a}, \tilde{u} : [0, 2L] \to \mathbf{R}$$

$$\tilde{a}(x) = \begin{cases} a(x), & x \in [0,L], \\ a(2L-x), & x \in (L, 2L] \end{cases}$$

$$\tilde{u}(x) = \begin{cases} u(x), & x \in [0,L], \\ -u(2L-x), & x \in (L, 2L]. \end{cases} \tag{2.103}$$

Then $\tilde{u} \in H^1(0, 2L)$ and we claim that $\tilde{u}$ is a (nontrivial) solution of

$$w''(x) + \tilde{a}(x)w(x) = 0, \ x \in (0, 2L), \ w'(0) = w'(2L) = 0. \tag{2.104}$$

To see this, we need to demonstrate

$$\int_0^{2L} \tilde{u}'(x)z'(x)\,dx = \int_0^{2L} \tilde{a}(x)\tilde{u}(x)z(x)\,dx, \quad \forall\, z \in H^1(0,2L). \tag{2.105}$$

If $z \in H^1(0,2L)$ satisfies

$$z(L) = 0 \tag{2.106}$$

then z, restricted to the interval $[0,L]$ is a test function for mixed problem (2.91) and therefore

$$\int_0^L \tilde{u}'(x)z'(x)\,dx = \int_0^L u'(x)z'(x)\,dx$$

$$= \int_0^L a(x)u(x)z(x)\,dx = \int_0^L \tilde{a}(x)\tilde{u}(x)z(x)\,dx. \tag{2.107}$$

Moreover, since function $z(2L - y), y \in [0,L]$, is also a test function for mixed problem (2.91), we have

$$\int_L^{2L} \tilde{u}'(x)z'(x)\,dx = \int_0^L u'(y)z'(2L - y)\,dy$$

$$= -\int_0^L a(y)u(y)z(2L - y)\,dy = \int_L^{2L} \tilde{a}(x)\tilde{u}(x)z(x)\,dx. \tag{2.108}$$

From (2.107) and (2.108) we deduce (2.105) when z satisfies (2.106). But in the interval $[0,2L]$, function $\tilde{a}(x)\tilde{u}(x)$ is an odd function with respect to L. This implies (2.105) when $z \equiv 1$. Finally, as any $z \in H^1(0,2L)$ may be written in the form $z(x) = (z(x) - z(L)) + z(L)$, we conclude (2.105).

Once we have proved that $\tilde{u}$ is a (nontrivial) solution of (2.104) associated with function $\tilde{a}$, we would need to have the sign condition

$$\int_0^{2L} \tilde{a}(x)\,dx \geq 0 \tag{2.109}$$

since this property is included into the definition of the set $\Lambda(0,2L)$. But

$$\int_0^{2L} \tilde{a}(x)\,dx = 2\int_0^L a(x)\,dx$$

and $a \in \Lambda^*(0,L)$, a set where no sign conditions is assumed. This difficulty may be overcome by using some eigenvalue ideas. In fact, we will prove

$$\forall\, a \in \Lambda^*(0,L),\ \exists\, k \in (0,1] : ka^+ \in \Lambda^*(0,L). \tag{2.110}$$

To establish this, note that if $a \in \Lambda^*(0, L)$ then $\int_0^L u'^2(x)\, dx = \int_0^L a(x)u^2(x)\, dx$
for some nontrivial function $u \in H^*$. Therefore the set $\{x \in (0, L) : a(x) > 0\}$ has
positive measure. As a consequence, the eigenvalue problem

$$w''(x) + \lambda a(x)w(x) = 0, \; x \in (0, L), \; w'(0) = w(L) = 0 \tag{2.111}$$

has a sequence of positive eigenvalues $\lambda_1(a) < \lambda_2(a) < \ldots$

Moreover, $a \in \Lambda^*(0, L)$ implies $\lambda_1(a) \leq 1$. Since $a^+ \geq a$, we have $\lambda_1(a^+) \leq \lambda_1(a)$. As $\lambda_1(a^+)a^+ \in \Lambda^*(0, L)$, this proves (2.110).

Now, from (2.110) we have $\|ka^+\|_{L^p(0,L)} \leq \|a^+\|_{L^p(0,L)}$. Therefore, it is clearly
not restrictive to assume from the beginning of the lemma we are proving that
$a(x) \geq 0$. This fact implies (2.109) and as a consequence, function $\tilde{a}$ defined
in (2.103) belongs to the set $\Lambda(0, 2L)$. Moreover, if $1 \leq p < \infty$,

$$2\|a\|_{L^p(0,L)}^p = \|\tilde{a}\|_{L^p(0,2L)}^p \geq \beta_p^p(0, 2L) = 2^{1-2p}\beta_p^p(0, L)$$

which imply

$$\|a\|_{L^p(0,L)} \geq \frac{1}{4}\beta_p(0, L),$$

for each function $a \in \Lambda^*(0, L) \bigcap L^p(0, L)$ such that $a(x) \geq 0$. From this and (2.110)
we obtain the conclusion of the lemma if $1 \leq p < \infty$. Similar ideas may be used in
the case $p = \infty$. This finishes also the proof of Theorem 2.7.

Remark 2.18. The proof of Theorem 2.7 that we have given here is based on an
appropriate change of variables, but it is possible to carry out a different approach
by using similar ideas to those contained in Sect. 2.1. In this way, some additional
results for $\beta_p^*(0, L)$ may be proved. For instance, $\beta_p^*(0, L)$ is attained if and only if
$1 < p \leq \infty$

Next, we present some results on the existence and uniqueness of solutions of
linear b.v.p.

$$u''(x) + a(x)u(x) = f(x), \; x \in (0, L), \; u'(0) = u'(L) = 0. \tag{2.112}$$

Previously, if $a \in L^1(c, d) \setminus \{0\}$, $\int_c^d a(x)\, dx \geq 0$ and $1 \leq p \leq \infty$, it may be
convenient to introduce hypothesis **(Hp)***.

Hypothesis (Hp)* It is established as:

1. $\|a^+\|_{L^1(c,d)} \leq \beta_1^*(c, d)$ if $p = 1$.
2. $a \in L^p(c, d)$, $\|a^+\|_{L^p(c,d)} < \beta_p^*(c, d)$.

Remark 2.19. Let us observe that if a function a satisfies hypothesis **(Hp)*** for
some p, $1 \leq p \leq \infty$, then the unique solution of the boundary problems

$$u''(x) + a(x)u(x) = 0, \; x \in (c, d), \; u'(c) = u(d) = 0 \tag{2.113}$$

and

$$u''(x) + a(x)u(x) = 0, \ x \in (c,d), \ u(c) = u'(d) = 0 \tag{2.114}$$

is the trivial one.

Theorem 2.8. *Let $a \in L^1(0,L) \setminus \{0\}$ with $\int_0^L a(x)\,dx \geq 0$, satisfying:*

For each $c \in (0,L)$ either hypothesis $(\mathbf{Hp})^*$ in the interval $(0,c)$ or hypothesis $(\mathbf{Hq})^*$ in the interval (c,L) (here, $p, q \in [1,\infty]$ may depend on c).

Then for each $f \in L^1(0,L)$, the boundary value problem (2.112) has a unique solution.

Proof. Since (2.112) is a linear problem, it is sufficient to see that the unique solution of the homogeneous problem

$$u''(x) + a(x)u(x) = 0, \ x \in (0,L), \ u'(0) = u'(L) = 0 \tag{2.115}$$

is the trivial one. Now, if (2.115) has some nontrivial solution u, it was shown at the beginning of this chapter that u must have a zero d in the interval $(0,L)$. In this case, both problems **PM(0,d)** and **PM(d,L)** have nontrivial solutions. But by using either hypothesis $(\mathbf{Hp})^*$ in $(0,d)$ or hypothesis $(\mathbf{Hq})^*$ in (d,L), we have a contradiction.

In concrete examples, it may be convenient to choose $p = q$, independent from $c \in (0,L)$. To this respect, the following proposition may be of interest.

Proposition 2.1. *Let $1 < p < \infty$ and $a \in L^p(0,L)$. Then the following statements are equivalent:*

$$\forall \, c \in (0,L), \ \text{either } \|a^+\|_{L^p(0,c)} < \beta_p^*(0,c) \ \text{or } \|a^+\|_{L^p(c,L)} < \beta_p^*(c,L) \tag{2.116}$$

$$\exists \, x_0 \in (0,L) : \ \|a^+\|_{L^p(0,x_0)} < \beta_p^*(0,x_0) \ \text{and } \|a^+\|_{L^p(x_0,L)} < \beta_p^*(x_0,L). \tag{2.117}$$

Proof. Let (2.116) be satisfied. Function $c^{2-\frac{1}{p}} \|a^+\|_{L^p(0,c)}$ is continuous and increasing in the interval $c \in (0,L)$ whereas function $(L-c)^{2-\frac{1}{p}} \|a^+\|_{L^p(c,L)}$ is continuous and decreasing in $c \in (0,L)$. Then, choose x_0 as a point in $(0,L)$ such that

$$x_0^{2-\frac{1}{p}} \|a^+\|_{L^p(0,x_0)} = (L-x_0)^{2-\frac{1}{p}} \|a^+\|_{L^p(x_0,L)}. \tag{2.118}$$

Since (2.116) is fulfilled $\forall \, c \in (0,L)$, it is true in particular for $c = x_0$. But taking into account (2.118) and the relation (2.99)

$$\beta_p^*(x_0,L) = \beta_p^*(0, L - x_0) = \left(\frac{L - x_0}{x_0}\right)^{\frac{1}{p}-2} \beta_p^*(0, x_0)$$

if x_0 is as in (2.118), both inequalities in (2.116) are really the same inequality and, moreover, they are identical to (2.117).

Reciprocally, if (2.117) is satisfied and $c \in (0, L)$, we can distinguish two cases: $c \in (0, x_0]$ and $c \in (x_0, L)$. In the first case, we have

$$\|a^+\|_{L^p(0,c)} \leq \|a^+\|_{L^p(0,x_0)} < \beta_p^*(0, x_0)$$

$$= \left(\frac{x_0}{L}\right)^{\frac{1}{p}-2} \beta_p^*(0, L) \leq \left(\frac{c}{L}\right)^{\frac{1}{p}-2} \beta_p^*(0, L) = \beta_p^*(0, c).$$

A similar reasoning is valid if $c \in (x_0, L)$.

Remark 2.20. An analogous result may be demonstrated for $p = 1$ by replacing strict inequalities in (2.116) and (2.117) with non-strict ones. If $p = \infty$, (2.117) implies (2.116). As a consequence, if (2.117) is satisfied for $p = \infty$, the unique solution of (2.115) is the trivial one. However, in this last case, a more precise condition may be obtained. This is shown in the next proposition.

Proposition 2.2. *If function a fulfills*

$$a \in L^\infty(0, L) \setminus \{0\}, \quad \int_0^L a \geq 0 \quad \text{and} \quad \exists \, x_0 \in (0, L):$$

$$\max\{x_0^2 \|a^+\|_{L^\infty(0,x_0)}, \ (L - x_0)^2 \|a^+\|_{L^\infty(x_0,L)}\} \leq \frac{\pi^2}{4} \qquad \textbf{(H)}$$

and, in addition, either a^+ is not the constant $\pi^2/4x_0^2$ in the interval $[0, x_0]$ or a^+ is not the constant $\pi^2/4(L - x_0)^2$ in the interval $[x_0, L]$, then for each $f \in L^1(0, L)$, the boundary value problem (2.112) has a unique solution.

Proof. To prove this proposition, take into account that $\beta_\infty^*(0, x_0) = \pi^2/4x_0^2$ and that $\beta_\infty^*(x_0, L) = \pi^2/4(L - x_0)^2$. Then, if $d \in (0, x_0)$ we have

$$\|a^+\|_{L^\infty(0,d)} \leq \|a^+\|_{L^\infty(0,x_0)} \leq \beta_\infty^*(0, x_0) < \beta_\infty^*(0, d)$$

Therefore, problem **PM(0,d)** has only the trivial solution. If $d \in (x_0, L)$ a similar reasoning is valid and we obtain that problem **PM(d,L)** has only the trivial solution. Finally, if $d = x_0$, we would have

$$\|a^+\|_{L^\infty(0,x_0)} \leq \beta_\infty^*(0, x_0), \quad \|a^+\|_{L^\infty(x_0,L)} \leq \beta_\infty^*(x_0, L).$$

But since, in addition, we have that either a^+ is not the constant $\pi^2/4x_0^2$ in the interval $[0, x_0]$ or a^+ is not the constant $\pi^2/4(L - x_0)^2$ in the interval $[x_0, L]$, we deduce that either problem **PM(0,x_0)** or problem **PM(x_0,L)** has only the trivial solution. This proves that (2.115) has only the trivial solution and therefore we have the desired conclusion.

In particular, if $x_0 = L/2$ in Proposition 2.2, we obtain the classical result related to the so-called nonuniform nonresonance conditions with respect to the first positive eigenvalue $\frac{\pi^2}{L^2}$ [24–26]. However, if for instance, $x_0 \in (0, L/2)$, it is allowed the equality $\|a^+\|_{L^\infty(0,x_0)} = \pi^2/4x_0^2$ (which is a quantity greater than $\frac{\pi^2}{L^2}$) as long as $\|a^+\|_{L^\infty(x_0,L)} < \pi^2/4(L - x_0)^2$.

Remark 2.21. Hypothesis (**H**) is optimal in the sense that if a^+ is the constant $\pi^2/4x_0^2$ in the interval $[0, x_0]$ and a^+ is the constant $\pi^2/4(L - x_0)^2$ in the interval $[x_0, L]$, then (2.115) has the $C^2[0, L]$ nontrivial solution:

$$u(x) = \begin{cases} \dfrac{-x_0}{L - x_0} \cos \dfrac{\pi x}{2x_0}, & \text{if } x \in (0, x_0), \\[4mm] \cos \dfrac{\pi(L - x)}{2(L - x_0)}, & \text{if } x \in (x_0, L), \end{cases}$$

Remark 2.22. By using the definition of β_p, it is clear that if for some p, with $1 \leq p < \infty$, function a satisfies

$$\|a^+\|_{L^p(0,L)} < \beta_p(0, L) \tag{2.119}$$

then the unique solution of (2.115) is the trivial one. It is easy to prove that (2.119) implies (2.116). In fact, if (2.116) is not true for some $c \in (0, L)$, taking into account (2.102) and Theorem 2.7 we obtain

$$\begin{aligned} \|a^+\|^p_{L^p(0,L)} &= \|a^+\|^p_{L^p(0,c)} + \|a^+\|^p_{L^p(c,L)} \\ &\geq (\beta_p^*(0, c))^p + (\beta_p^*(c, L))^p \\ &= \left[\left(\frac{c}{L}\right)^{1-2p} + \left(\frac{L - c}{L}\right)^{1-2p} \right] \frac{(\beta_p(0, L))^p}{4^p} \geq (\beta_p(0, L))^p \end{aligned}$$

which is a contradiction with (2.119).

Previous remark shows that if we want to have a criterion implying that (2.115) has only the trivial solution, then (2.116) is better than (2.119). In order to prove that (2.116) is a strict generalization of (2.119), we show the following example (see [5] for more details).

Example. Let c_1, c_2 be two positive numbers and let us consider the two step potential

$$a(x) = \begin{cases} c_1^2, & \text{if } 0 \leq x < \dfrac{Lc_2}{c_1 + c_2}, \\[4mm] c_2^2, & \text{if } \dfrac{Lc_2}{c_1 + c_2} \leq x \leq L. \end{cases} \tag{2.120}$$

Then, for each p, $1 \leq p \leq \infty$, there exist c_1, c_2 such that

function a satisfies (2.117) (and therefore (2.116)) for $x_0 = \dfrac{Lc_2}{c_1 + c_2}$ $\qquad$ (2.121)

and

$$\|a^+\|_{L^q(0,L)} > \beta_q(0,L), \ \forall q \in [1,\infty]. \tag{2.122}$$

Let us remark that from (2.122) we cannot deduce that (2.115) has only the trivial solution. However, we can affirm that this fact is true from (2.121).

We finish this section with some results on the existence and uniqueness of solutions of nonlinear b.v.p.

$$u''(x) + f(x, u(x)) = 0, \ x \in (0,L), \ u'(0) = u'(L) = 0. \tag{2.123}$$

Taking into account the previous discussion, next theorem is a strict generalization of Theorem 3.1 in [3], Theorem B in [18] and (for ordinary differential equations) Theorem 7.1 in [4] and Theorem 2 in [26]. The proof, which is similar to the one given in [3, 4], combines the linear results of this section with Schauder's fixed point theorem. We omit the details.

Theorem 2.9. *Let us consider (2.123) where the following requirements are supposed:*

1. f and f_u are Caratheodory functions on $[0,L] \times \mathbf{R}$ and $f(\cdot,0) \in L^1(0,L)$.
2. There exist functions $\alpha, \beta \in L^\infty(0,L)$, satisfying

$$\alpha(x) \le f_u(x,u) \le \beta(x)$$

on $[0,L] \times \mathbf{R}$ and β satisfies for each $c \in (0,c)$ either hypothesis $(\mathbf{Hp})^*$ in the interval $(0,c)$ (for some $p \in [1,\infty]$), or hypothesis $(\mathbf{Hq})^*$ in the interval (c,L) (for some $q \in [1,\infty]$).
3. Moreover, we assume one of the following conditions:

  a.

$$\int_\Omega \alpha \ge 0, \ \alpha \not\equiv 0$$

  b.

$$\alpha \equiv 0, \ \exists s_0 \in \mathbf{R} \ s.t. \ \int_\Omega f(x,s_0) \, dx = 0, \ and f_u(x,u(x)) \not\equiv 0, \ \forall u \in C(\overline{\Omega}).$$

Then, problem (2.123) has a unique solution.

Remark 2.23. The idea of using qualitative properties of the mixed problem **PM(0,c)** in the study of resonant nonlinear problems like (2.123) has been previously employed by different authors. The interested reader may consult [13, 25, 27, 29] for the case where the nonlinearity f is restricted in one direction.

References

1. Borg, G.: On a Lyapunov criterion of stability. Am. J. Math. **71**, 67–70 (1949)
2. Brezis, H.: Analyse Fonctionnelle. Masson, Paris (1983)
3. Cañada, A., Montero, J.A., Villegas, S.: Lyapunov type inequalities and Neumann boundary value problems at resonance. Math. Inequal. Appl. **8**, 459–475 (2005)
4. Cañada, A., Montero, J.A., Villegas, S.: Lyapunov inequalities for partial differential equations. J. Funct. Anal. **237**, 176–193 (2006)
5. Cañada, A., Villegas, S.: Optimal Lyapunov inequalities for disfocality and Neumann boundary conditions using L^p norms. Discrete Contin. Dyn. Syst. Ser. A **20**, 877–888 (2008)
6. Cañada, A., Villegas, S.: Stability, resonance and Lyapunov inequalities for periodic conservative systems. Nonlinear Anal. **74**, 1913–1925 (2011)
7. Choquet, G.: Topology. Academic, New York (1966)
8. Courant, R., Hilbert, D.: Methods of Mathematical Physics. Wiley Interscience, New York (1962)
9. Croce, G., Dacorogna, B.: On a generalized Wirtinger inequality. Discrete Contin. Dyn. Syst. **9**, 1329–1341 (2003)
10. Dacorogna, B.: Introduction to the Calculus of Variations. Imperial College Press, London (2004)
11. Dacorogna, B., Gangbo, W., Subía, N.: Sur une généralisation de l'inégalité de Wirtinger. Ann. Inst. Henri Poincaré Anal. Non Linéaire **9**, 29–50 (1992)
12. Deimling, K.: Nonlinear Functional Analysis. Springer, Berlin (1985)
13. Dong, Y.: A Neumann problem at resonance with the nonlinearity restricted in one direction. Nonlinear Anal. **51**, 739–747 (2002)
14. Drábek, P.: Nonlinear eigenvalue problems and Fredholm alternative. In: Drábek, P., Krejči, P., Takáč, P. (eds.) Nonlinear Differential Equations. Research Notes in Mathematics Series, vol. 404, pp. 1–46. Chapman and Hall/CRC, London (1999)
15. Harris, B.J.: On an inequality of Lyapunov for disfocality. J. Math. Anal. Appl. **146**, 495–500 (1990)
16. Hartman, P.: Ordinary Differential Equations. Wiley, New York (1964)
17. Huaizhong, W., Yong, L.: Existence and uniqueness of periodic solutions for Duffing equations across many points of resonance. J. Differ. Equ. **108**, 152–169 (1994)
18. Huaizhong, W., Yong, L.: Neumann boundary value problems for second-order ordinary differential equations across resonance. SIAM J. Control Optim. **33**, 1312–1325 (1995)
19. Kwong, M.K.: On Lyapunov's inequality for disfocality. J. Math. Anal. Appl. **83**, 486–494 (1981)
20. Landesman, E.M., Lazer, A.C.: Linear eigenvalues and a nonlinear boundary value problem. Pac. J. Math. **33**, 311–328 (1970)
21. López, G., Montero, J.A.: Second order Neumann boundary value problems across resonance. ESAIM Control Optim. Calc. Var. **12**, 398–408 (2006)
22. Lyapunov, M.A.: Problème général de la stabilité du mouvement. Ann. Fac. Sci. Univ. Tolouse Sci. Mat. Sci. Phys. **9**, 203–474 (1907)
23. Magnus, W., Winkler, S.: Hill's Equation. Dover, New York (1979)
24. Mawhin, J., Ward, J.R.: Nonuniform nonresonance conditions at the two first eigenvalues for periodic solutions of forced Liénard and Duffing equations. Rocky Mountain J. Math. **12**, 643–654 (1982)
25. Mawhin, J., Ward, J.R.: Periodic solutions of some forced Liénard differential equations at resonance. Arch. Math. (Basel) **41**, 337–351 (1983)
26. Mawhin, J., Ward, J.R., Willem, M.: Variational methods and semilinear elliptic equations. Arch. Ration. Mech. Anal. **95**, 269–277 (1986)
27. Mawhin, J., Ruiz, D.: A strongly nonlinear Neumann problem at resonance with restrictions on the nonlinearity just in one direction. Topol. Methods Nonlinear Anal. **20**, 1–14 (2002)

28. Talenti, G.: Best constant in Sobolev inequality. Ann. Mat. Pura Appl. **110**, 353–372 (1976)
29. Villegas, S.: A Neumann problem with asymmetric nonlinearity and a related minimizing problem. J. Differ. Equ. **145**, 145–155 (1998)
30. Zhang, M.: Certain classes of potentials for p-Laplacian to be non-degenerate. Math. Nachr. **278**, 1823–1836 (2005)

Chapter 3
Higher Eigenvalues

Abstract This chapter is devoted to the study of L_1 Lyapunov-type inequalities for different boundary conditions at higher eigenvalues. Our main result is derived from a detailed analysis about the number and distribution of zeros of nontrivial solutions and their first derivatives, together with the use of suitable minimization problems. As in the classical result by Lyapunov at the first eigenvalue, the L_1 best constant at higher eigenvalues is not attained. The linear study on periodic and antiperiodic boundary conditions is used to establish some new conditions for the stability of linear periodic equations. Moreover, we use the Schauder fixed point theorem to provide new conditions about the existence and uniqueness of solutions for resonant nonlinear problems at higher eigenvalues.

3.1 Motivation of the Problem: Neumann and Dirichlet Boundary Conditions

From Corollary 2.1, it is easily proved that the problem

$$u''(x) + a(x)u(x) = 0, \ x \in (0, L), \ u'(0) = u'(L) = 0 \tag{3.1}$$

has only the trivial solution if function a satisfies

$$a \in L^\infty(0, L) \setminus \{0\}, \ \int_0^L a \geq 0, \ a^+ \prec \pi^2/L^2, \tag{3.2}$$

where, remember that for $c, d \in L^1(0, L)$, we write $c \prec d$ if $c(x) \leq d(x)$ for a.e. $x \in [0, L]$ and $c(x) < d(x)$ on a set of positive measure. In particular, (3.2) is satisfied if

$$0 \prec a \prec \pi^2/L^2. \tag{3.3}$$

This is a very well-known result which provides a nonuniform nonresonance condition with respect to the two first eigenvalues $\lambda_0 = 0$ and $\lambda_1 = \pi^2/L^2$ of the eigenvalue problem

$$u''(x) + \lambda u(x) = 0, \ x \in (0, L), \ u'(0) = u'(L) = 0 \tag{3.4}$$

© The Author(s) 2015
A. Cañada, S. Villegas, *A Variational Approach to Lyapunov Type Inequalities*,
SpringerBriefs in Mathematics, DOI 10.1007/978-3-319-25289-6_3

(see [15–17]). From this point of view, it may be affirmed that the nonuniform nonresonance condition (3.3) is in fact an $L_\infty - L_\infty$ Lyapunov inequality at the two first eigenvalues λ_0 and λ_1.

On the other hand, the set of eigenvalues of (3.4) is given by $\lambda_n = n^2\pi^2/L^2$, $n \in \mathbb{N} \cup \{0\}$ and by using a general result due to Dolph [7] it can be proved that, if for some $n \geq 0$ function a satisfies

$$\lambda_n \prec a \prec \lambda_{n+1} \tag{3.5}$$

then (3.1) has only the trivial solution (see [14], Lemma 2.1, for some generalizations of (3.5) to more general boundary value problems).

It is clear that, for $n \geq 1$, condition (3.5) cannot be obtained from the results of Sect. 2.1, which are concerning to the two first eigenvalues.

These observations motivate the contents of this chapter where, for any given natural number $n \geq 1$ and function a satisfying the L_∞ restriction $\lambda_n \prec a$, we obtain an L_1 Lyapunov inequality. Under the restriction $\lambda_n \prec a$, the L_∞ Lyapunov inequality is exactly (3.5) and in this sense, it is natural to say that this chapter deals with $L_\infty - L_1$ Lyapunov inequalities at higher eigenvalues . In particular we prove, as it happens in the classical Lyapunov inequality at the first eigenvalue, that the best constant is not attained for any value of n. To the best of our knowledge, the L_p case with $1 < p < \infty$ has not been solved yet.

Next, we enunciate and prove the main result of this section.

If $n \in \mathbb{N}$ is fixed, we introduce the set Λ_n as

$$\Lambda_n = \{a \in L^1(0,L) : \lambda_n \prec a \text{ and (3.1) has nontrivial solutions }\} \tag{3.6}$$

and the constant $\beta_{1,n}$ as

$$\beta_{1,n} \equiv \inf_{a \in \Lambda_n} \|a\|_{L^1(0,L)} \tag{3.7}$$

Theorem 3.1.

$$\beta_{1,n} = \lambda_n L + \frac{2\pi n(n+1)}{L} \cot \frac{\pi n}{2(n+1)} = n^2\pi^2/L + \frac{2\pi n(n+1)}{L} \cot \frac{\pi n}{2(n+1)} \tag{3.8}$$

and $\beta_{1,n}$ is not attained.

Proof. The proof will be carried out into several steps:

1. Analysis on the number and distribution of zeros of nontrivial solutions u of Eq. (3.1) and their first derivatives u'

In this step, we include an optimal estimation about the corresponding distances between the zeros of u and u'. To this respect, we compare our problem with other one with mixed boundary conditions. This allows us to obtain a more precise information than that obtained when the classical Sturm separation theorem is used, where the problem is compared with the case of Dirichlet boundary conditions ([8]).

Since $a \in \Lambda_n$, it is clear that between two consecutive zeros of the function u there must exist a zero of the function u' and between two consecutive zeros of the function u' there must exist a zero of the function u.

Lemma 3.1. *Let $a \in \Lambda_n$ be given and u any nontrivial solution of (3.1). If the zeros of u' in $[0, L]$ are denoted by $0 = x_0 < x_2 < \ldots < x_{2m} = L$ and the zeros of u in $(0, L)$ are denoted by $x_1 < x_3 < \ldots < x_{2m-1}$, then:*

1. *$x_{i+1} - x_i \leq \frac{L}{2n}, \ \forall\, i : 0 \leq i \leq 2m-1$. Moreover, at least one of these inequalities is strict.*
2. *$m \geq n + 1$. Moreover, any value $m \geq n + 1$ is possible.*

Proof. Let i, $0 \leq i \leq 2m-1$, be given. Then, function u satisfies either the problem

$$u''(x) + a(x)u(x) = 0, \ x \in (x_i, x_{i+1}), \ u(x_i) = 0, \ u'(x_{i+1}) = 0 \tag{3.9}$$

or the problem

$$u''(x) + a(x)u(x) = 0, \ x \in (x_i, x_{i+1}), \ u'(x_i) = 0, \ u(x_{i+1}) = 0 \tag{3.10}$$

Let us assume the first case. The reasoning in the second case is similar. Note that u may be chosen such that $u(x) > 0, \ \forall\, x \in (x_i, x_{i+1})$. Let us denote by μ_1^i and φ_1^i, respectively, the principal eigenvalue and eigenfunction of the eigenvalue problem

$$v''(x) + \mu v(x) = 0, \ x \in (x_i, x_{i+1}), \ v(x_i) = 0, \ v'(x_{i+1}) = 0 \tag{3.11}$$

It is known that

$$\mu_1^i = \frac{\pi^2}{4(x_{i+1} - x_i)^2}, \quad \varphi_1^i(x) = \sin\frac{\pi(x - x_i)}{2(x_{i+1} - x_i)}. \tag{3.12}$$

Choosing φ_1^i as test function in the weak formulation of (3.9) and u as test function in the weak formulation of (3.11) for $\mu = \mu_1^i$ and $v = \varphi_1^i$, we obtain

$$\int_{x_i}^{x_{i+1}} (a(x) - \mu_1^i) u \varphi_1^i(x)\, dx = 0. \tag{3.13}$$

Then, if $x_{i+1} - x_i > \frac{L}{2n}$, we have

$$\mu_1^i = \frac{\pi^2 L^2}{4(x_{i+1} - x_i)^2 L^2} < \frac{n^2 \pi^2}{L^2} = \lambda_n \leq a(x), \text{ a.e. in } (x_i, x_{i+1})$$

which is a contradiction with (3.13). Consequently, $x_{i+1} - x_i \leq \frac{L}{2n}, \forall\, i : 0 \leq i \leq 2m - 1$. Also, since $\lambda_n \prec a$ in the interval $(0, L)$, we must have $\lambda_n \prec a$ in some subinterval (x_j, x_{j+1}). If $x_{j+1} - x_j = \frac{L}{2n}$, it follows $\mu_1^j \prec a$ in (x_j, x_{j+1}) and this is again a contradiction with (3.13). These reasonings complete the first part of the lemma. For the second one, let us observe that

$$L = \sum_{i=0}^{2m-1} (x_{i+1} - x_i) < 2m\frac{L}{2n}.$$

In consequence, $m > n$. Also, note that for any given natural number $q \geq n + 1$, function $a(x) \equiv \lambda_q$ belongs to Λ_n and for function $u(x) = \cos\frac{q\pi x}{L}$, we have $m = q$.

2. **Study of some associated minimization problems.** If i, with $0 \leq i \leq 2m - 1$ is given and u satisfies (3.9), then

$$\int_{x_i}^{x_{i+1}} u'^2(x) = \int_{x_i}^{x_{i+1}} a(x)u^2(x)$$

$$= \int_{x_i}^{x_{i+1}} (a(x) - \lambda_n)u^2(x) + \int_{x_i}^{x_{i+1}} \lambda_n u^2(x).$$

Therefore,

$$\int_{x_i}^{x_{i+1}} u'^2(x) - \lambda_n \int_{x_i}^{x_{i+1}} u^2(x) \leq \|a - \lambda_n\|_{L^1(x_i,x_{i+1})} \|u^2\|_{L^\infty(x_i,x_{i+1})}.$$

Since u' has no zeros in the interval (x_i, x_{i+1}) and $u(x_i) = 0$, we have $\|u^2\|_{L^\infty(x_i,x_{i+1})} = u^2(x_{i+1})$ and consequently

$$\|a - \lambda_n\|_{L^1(x_i,x_{i+1})} \geq \frac{\int_{x_i}^{x_{i+1}} u'^2 - \lambda_n \int_{x_i}^{x_{i+1}} u^2}{u^2(x_{i+1})}, \quad \text{if } i \text{ is odd.} \tag{3.14}$$

Analogously, if u satisfies (3.10), we deduce

$$\|a - \lambda_n\|_{L^1(x_i,x_{i+1})} \geq \frac{\int_{x_i}^{x_{i+1}} u'^2 - \lambda_n \int_{x_i}^{x_{i+1}} u^2}{u^2(x_i)}, \quad \text{if } i \text{ is even.} \tag{3.15}$$

The inequalities (3.14) and (3.15) motivate the next Lemma. The proof uses classical and standard methods of the calculus of variations.

Lemma 3.2. *Assume that $a < b$ and $0 < M \leq \dfrac{\pi^2}{4(b-a)^2}$ are given real numbers. Let $H = \{u \in H^1(a,b) : u(a) = 0, u(b) \neq 0\}$. If $J : H \rightarrow \mathbf{R}$ is defined by*

$$J(u) = \frac{\int_a^b u'^2 - M \int_a^b u^2}{u^2(b)} \tag{3.16}$$

and $m \equiv \inf_{u \in H} J(u)$, m is attained and

$$m = M^{1/2} \cot(M^{1/2}(b - a)). \tag{3.17}$$

Moreover, if $u \in H$, then $J(u) = m \iff u(x) = k \dfrac{\sin(M^{1/2}(x - a))}{\sin(M^{1/2}(b - a))}$ for some nonzero constant k.

Proof. The constant $\delta_1 = \dfrac{\pi^2}{4(b - a)^2}$ is the principal eigenvalue of the eigenvalue problem $v''(x) + \delta v(x) = 0$, $v(a) = 0$, $v'(b) = 0$ with associated eigenfunction $w(x) = \sin \dfrac{\pi(x - a)}{2(b - a)}$. Therefore, if $M = \dfrac{\pi^2}{4(b - a)^2}$, $m = 0$ and it is attained at function w.

If $M < \delta_1 = \frac{\pi^2}{4(b-a)^2}$, there exists some positive constant c such that

$$\int_a^b u'^2 - M \int_a^b u^2 \geq c \int_a^b u'^2, \ \forall u \in H. \tag{3.18}$$

If $\{u_n\} \subset H$ is a minimizing sequence for J, since the sequence $\{k_n u_n\}$, $k_n \neq 0$, is also a minimizing sequence, we can assume without loss of generality that $u_n(b) = 1$. From (3.18) we deduce that $\int_a^b u_n'^2$ is bounded. So, we can suppose, up to a subsequence, that $u_n \rightharpoonup u_0$ in $H^1(a, b)$ and $u_n \to u_0$ in $C[a, b]$ (with the uniform norm). The strong convergence in $C[a, b]$ gives us $u_0(b) = 1$. The weak convergence in H implies $J(u_0) \leq \liminf J(u_n) = m$. Then u_0 is a minimizer.

Since $J(u_0) = \min\{J(v) : v \in H^1(a, b), v(a) = 0, v(b) = 1\}$, Lagrange multiplier Theorem implies that there are real numbers α_1, α_2 such that

$$2 \int_a^b u_0' v' - 2M \int_a^b u_0 v - \alpha_1 v(b) - \alpha_2 v(a) = 0, \ \forall v \in H^1(a, b).$$

In particular,

$$\int_a^b u_0' v' - M \int_a^b u_0 v = 0, \ \forall v \in H^1(a, b) : v(a) = v(b) = 0.$$

We conclude that u_0 satisfies the problem

$$u_0''(x) + M u_0(x) = 0, \ x \in (a, b), \ u_0(a) = 0, \ u_0(b) = 1. \tag{3.19}$$

Note that since $M < \dfrac{\pi^2}{(b - a)^2}$, (3.19) has a unique solution, which is given by

$$u_0(x) = \frac{\sin(M^{1/2}(x - a))}{\sin(M^{1/2}(b - a))}. \tag{3.20}$$

Finally, a rudimentary calculation gives $J(u_0) = M^{1/2} \cot(M^{1/2}(b - a))$. This proves the lemma.

Now, we combine the inequalities (3.14), (3.15), and the previous Lemma to conclude that if $a \in \Lambda_n$ is given and u is any nontrivial solution of (3.1), then

$$\|a - \lambda_n\|_{L^1(0,L)} \geq \frac{n\pi}{L} \sum_{i=0}^{2m-1} \cot\left(\frac{n\pi}{L}(x_{i+1} - x_i)\right),\qquad(3.21)$$

where the zeros of u' are denoted by $0 = x_0 < x_2 < \ldots < x_{2m} = L$ and the zeros of u are denoted by $x_1 < x_3 < \ldots < x_{2m-1}$. This gives rise to the following Lemma, where an *elementary minimization problem* is studied.

Lemma 3.3. *Given any $r \in \mathbf{N}$ and $S \in \mathbf{R}^+$ satisfying $r\pi > 2S$, let*

$$Z = \left\{ z = (z_0, z_1, \ldots, z_{r-1}) \in (0, \pi/2]^r : \sum_{i=0}^{r-1} z_i = S \right\}.$$

If $F : Z \to \mathbf{R}$ is defined by

$$F(z) = \sum_{i=0}^{r-1} \cot z_i$$

then $\inf_{z \in Z} F(z)$ is attained and its value is $r \cot \frac{S}{r}$. Moreover, $z \in Z$ is a minimizer if and only if $z_i = \frac{S}{r}$, $\forall\, 0 \leq i \leq r - 1$.

Proof. Let us observe that $\forall\, z \in Z$, $\cot z_i \geq 0$, $0 \leq i \leq r-1$. Moreover, if $z_i \to 0^+$ for some $0 \leq i \leq r - 1$, then $\cot z_i \to +\infty$. Also, since $r\pi > 2S$, if $z \in Z$ is such that $z_i = \pi/2$, for some $0 \leq i \leq r - 1$, then there must exist some $0 \leq j \leq r - 1$ such that $z_j < \pi/2$. Let us choose the point $z^* \in Z$ defined (for $\delta > 0$ sufficiently small) as $z_k^* = z_k$, if $k \neq i$ and $k \neq j$, $z_i^* = \frac{\pi}{2} - \delta$, $z_j^* = z_j + \delta$. An elementary calculation shows

$$F(z^*) - F(z) = \frac{\cot z_j(1 - \cot z_j \cot \delta)}{\cot \delta(\cot z_j + \cot \delta)}$$

which is a negative number for δ sufficiently small. Consequently, there exits a sufficiently small positive constant ε_1 such that

$$\inf_{z \in Z} F(z) = \min_{z \in [\varepsilon_1, \frac{\pi}{2}]^r} F(z) = \min_{z \in (\varepsilon_1, \frac{\pi}{2})^r} F(z).$$

Then, if $z \in Z$ is any minimizer of F, Lagrange multiplier Theorem implies that there is $\lambda \in \mathbf{R}$ such that

$$\frac{-1}{\sin^2 z_i} + \lambda = 0,\ 0 \leq i \leq r - 1,\ \sum_{i=0}^{r-1} z_i = S.$$

We conclude $z_i = \frac{S}{r}$, $0 \leq i \leq r - 1$ and the lemma is proved.

From two previous lemmas, we obtain a lower bound for $\beta_{1,n}$ (this lower bound will be, ultimately, the infimum defined in (3.7)):

$$\beta_{1,n} \geq \lambda_n L + \frac{n\pi}{L} 2(n+1) \cot \frac{n\pi}{2(n+1)}. \qquad (3.22)$$

In fact, if $a \in \Lambda_n$ is given and u is any nontrivial solution of (3.1) (again, the zeros of u' are denoted by $0 = x_0 < x_2 < \ldots < x_{2m} = L$ and the zeros of u are denoted by $x_1 < x_3 < \ldots < x_{2m-1}$), then we obtain from Lemmas 3.21 and 3.3 (with $r = 2m$, $S = n\pi$, and $z_i = \frac{n\pi}{L}(x_{i+1} - x_i)$)

$$\|a - \lambda_n\|_{L^1(0,L)} \geq \frac{n\pi}{L} \sum_{i=0}^{2m-1} \cot\left(\frac{n\pi}{L}(x_{i+1} - x_i)\right) \geq \frac{n\pi}{L} 2m \cot \frac{n\pi}{2m}. \qquad (3.23)$$

Finally, taking into account that the function $2m \cot \frac{n\pi}{2m}$ is strictly increasing with respect to m, and that $m \geq n+1$, we deduce (3.22).

3. A minimizing sequence for $\beta_{1,n}$ (and $\beta_{1,n}$ is not attained)

It is possible to define a minimizing sequence for $\beta_{1,n}$ which shows that the lower bound found in (3.22) is, in fact, the infimum defined in (3.7). The relations (2.35) motivate the definition of this minimizing sequence. Anyway, this is a very technical subject and the lector is referred to [5] for further details.

A minimizing sequence can be defined in the following way:

Let $\varepsilon > 0$ be sufficiently small. Let us define the function $u_\varepsilon : [0, L] \to \mathbf{R}$ by

$$u_\varepsilon(x) = \begin{cases} -\sin(\frac{n\pi}{L}(x - \frac{L}{2(n+1)})) + \frac{n\pi}{L}\frac{(x-\varepsilon)^3}{3\varepsilon^2} \cos(\frac{n\pi}{2(n+1)}), & \text{if } 0 \leq x \leq \varepsilon, \\[2mm] -\sin(\frac{n\pi}{L}(x - \frac{L}{2(n+1)})), & \text{if } \varepsilon \leq x \leq \frac{L}{2(n+1)}, \\[2mm] -u_\varepsilon(\frac{2L}{2(n+1)} - x), & \text{if } \frac{L}{2(n+1)} \leq x \leq \frac{2L}{2(n+1)}, \\[2mm] u_\varepsilon(\frac{4L}{2(n+1)} - x), & \text{if } \frac{2L}{2(n+1)} \leq x \leq \frac{4L}{2(n+1)}, \\[2mm] -u_\varepsilon(\frac{6L}{2(n+1)} - x), & \text{if } \frac{4L}{2(n+1)} \leq x \leq \frac{6L}{2(n+1)}, \\[2mm] \ldots \end{cases} \qquad (3.24)$$

Then $u_\varepsilon \in C^2[0, L]$, the function $a_\varepsilon(x) \equiv \frac{-u_\varepsilon''(x)}{u_\varepsilon(x)}$, $\forall x \in [0, L]$, $x \neq \frac{(2k-1)L}{2(n+1)}$, $1 \leq k \leq n+1$, belongs to Λ_n and

$$\liminf_{\varepsilon \to 0^+} \|a_\varepsilon - \lambda_n\|_{L^1(0,L)} = \frac{n\pi}{L} 2(n+1) \cot \frac{n\pi}{2(n+1)}. \qquad (3.25)$$

Lastly, it can be proved that $\beta_{1,n}$ is not attained. To see this, let $a \in \Lambda_n$ be such that $\|a - \lambda_n\|_{L^1(0,L)} = \beta_{1,n}$ and let u be any nontrivial solution of (3.1) associated with the function a. Once more, we denote the zeros of u' by $0 = x_0 < x_2 < \ldots < x_{2m} = L$ and the zeros of u by $x_1 < x_3 < \ldots < x_{2m-1}$. Then,

$$
\begin{aligned}
\beta_{1,n} = \|a - \lambda_n\|_{L^1(0,L)} &= \sum_{i=0}^{2m-1} \|a - \lambda_n\|_{L^1(x_i, x_{i+1})} \\
&\geq \sum_{i=0}^{2m-1} J_i(u) \geq \frac{n\pi}{L} \sum_{i=0}^{2m-1} \cot \frac{n\pi(x_{i+1} - x_i)}{L} \\
&\geq \frac{n\pi}{L} 2m \cot \frac{n\pi}{2m} \geq \frac{n\pi}{L} 2(n+1) \cot \frac{n\pi}{2(n+1)} = \beta_{1,n},
\end{aligned}
\qquad (3.26)
$$

where $J_i(u)$ is given by

$$
J_i(u) = \frac{\displaystyle\int_{x_i}^{x_{i+1}} u'^2 - \lambda_n \int_{x_i}^{x_{i+1}} u^2}{u^2(x_{i+1})}, \quad \text{if } u(x_i) = 0
$$

and by

$$
J_i(u) = \frac{\displaystyle\int_{x_i}^{x_{i+1}} u'^2 - \lambda_n \int_{x_i}^{x_{i+1}} u^2}{u^2(x_i)}, \quad \text{if } u(x_{i+1}) = 0.
$$

Consequently, all inequalities in (3.26) transform into equalities. In particular we obtain from Lemma 3.3

$$
m = n + 1, \quad x_{i+1} - x_i = \frac{L}{2(n+1)}, \quad 0 \leq i \leq 2n + 1.
$$

Also, it follows

$$
J_i(u) = \frac{n\pi}{L} \cot \frac{n\pi}{L} \frac{L}{2(n+1)}, \quad 0 \leq i \leq 2n + 1.
$$

From Lemma 3.2 we deduce that, up to some nonzero constants, function u fulfills in each interval $[x_i, x_{i+1}]$

$$
u(x) = \frac{\sin \frac{n\pi}{L}(x - x_i)}{\sin \frac{n\pi}{L}(x_{i+1} - x_i)}, \quad \text{if } i \text{ is odd}
$$

$$
u(x) = \frac{\sin \frac{n\pi}{L}(x - x_{i+1})}{\sin \frac{n\pi}{L}(x_i - x_{i+1})}, \quad \text{if } i \text{ is even.}
$$

In particular, in the interval $[0, \frac{L}{2(n+1)}]$, u must be the function

$$u(x) = \frac{\sin \frac{n\pi}{L}\left(x - \frac{L}{2(n+1)}\right)}{\sin \frac{n\pi}{L}\left(-\frac{L}{2(n+1)}\right)}$$

which does not satisfy the condition $u'(0) = 0$. The conclusion is that $\beta_{1,n}$ is not attained.

Remark 3.1. In the previous reasonings, it is not necessary that n be a natural number. In fact, a similar study can be carried out if $n \in \mathbf{R}^+ \setminus \mathbf{N}$. The conclusions on the number of zeros, etc. of the nontrivial solutions u of (3.1) and its derivative u' are as previously if one substitutes n by the smallest natural number greater than n. Then, $\beta_{1,n}$ can be considered as a function of $n \in (0, +\infty)$ and $\lim_{n\to 0+} \beta_{1,n} = \frac{4}{L}$, the constant of the classical L^1 Lyapunov inequality at the first eigenvalue.

Remark 3.2. The case where $L = 1$ and function a satisfies the condition $A \leq a(x) \leq B$, a.e. in $(0, L)$ where $\lambda_k < A < \lambda_{k+1} \leq B$ for some $k \in \mathbf{N} \cup \{0\}$, has been examined in [20], where the authors use Optimal Control theory methods . In this paper, the authors define the set $\Lambda_{A,B}$ as the set of functions $a \in L^1(0, 1)$ such that $A \leq a(x) \leq B$, a.e. in $(0, 1)$ and (3.1) has nontrivial solutions. Then, by using the Pontryagin's maximum principle they prove that the number

$$\beta_{A,B} \equiv \inf_{a\in\Lambda_{A,B}} \|a\|_{L^1(0,1)}$$

is attained. In addition, they calculate $\lim_{B\to+\infty} \beta_{A,B}$.

Remark 3.3. We can use our methods to do an analogous study for other boundary conditions. For example, Dirichlet boundary value problems, where there is no change with respect to Neumann ones (see [9, 21] for a different treatment of the problem, where Optimal Control methods are used).

Because of its importance in stability theory, in the next section we deal with the cases of periodic and antiperiodic boundary conditions. Only those innovative details with respect to the Neumann case are shown (see [6] for a more detailed discussion).

3.2 Periodic and Antiperiodic Boundary Conditions

The first part of this section deals with L_1-Lyapunov inequality for the periodic problem

$$u''(t) + a(t)u(t) = 0, \ t \in (0, T), \ u(0) - u(T) = u'(0) - u'(T) = 0 \qquad (3.27)$$

at higher eigenvalues, when $a \in L_T(\mathbf{R}, \mathbf{R})$ (remember, the set of T-periodic functions $a : \mathbf{R} \to \mathbf{R}$ such that $a|_{[0,T]} \in L^1(0, T)$).

The set of eigenvalues of

$$u''(t) + \lambda u(t) = 0, \ t \in (0, T), \ u(0) - u(T) = u'(0) - u'(T) = 0 \qquad (3.28)$$

is given by $\lambda_0 = 0$, $\lambda_{2n-1} = \lambda_{2n} = (2n)^2\pi^2/T^2$, $n \in \mathbf{N}$ and if $n \in \mathbf{N}$ is fixed, we introduce the set Λ_n^{per} as

$$\Lambda_n^{\text{per}} = \{a \in L_T(\mathbf{R}, \mathbf{R}) : \lambda_{2n-1} \prec a \text{ and (3.27) has nontrivial solutions } \} \qquad (3.29)$$

We define

$$\beta_{1,n}^{\text{per}} = \inf_{a \in \Lambda_n^{\text{per}}} \|a\|_{L^1(0,T)}. \qquad (3.30)$$

As in the case of Neumann boundary conditions, the main result is derived from a detailed analysis on the number and distribution of zeros of nontrivial solutions of (3.27) and their first derivatives, together with the study of some special minimization problems. Since the proof here is similar to that used in the Neumann case, the interested reader can consult [6] for more details.

It must be remarked that if $a \in \Lambda_n^{\text{per}}$, and u is any nontrivial solution of (3.27), then u is not a constant function. In addition, u must have a zero in the interval $[0, T]$. If $r \in [0, T]$ is such that $u(r) = 0$, the periodic and nontrivial function $v(t) = u(r + t)$ satisfies $v''(t) + a(r + t)v(t) = 0$, $t \in (0, T)$ and $\|a(r + \cdot) - \lambda_{2n-1}\|_{L^1(0,T)} = \|a(\cdot) - \lambda_{2n-1}\|_{L^1(0,T)}$. Finally, since $a \in \Lambda_n^{\text{per}}$, $n \in \mathbf{N}$, it is clear that between two consecutive zeros of the function u there must exist a zero of the function u' and between two consecutive zeros of the function u' there must exist a zero of the function u. By using the same ideas as in Theorem 3.1, if u is any nontrivial solution of (3.27) such that $u(0) = u(T) = 0$ and the zeros of u in $[0, T]$ are denoted by $0 = t_0 < t_2 < \ldots < t_{2m} = T$ and the zeros of u' in $(0, T)$ are denoted by $t_1 < t_3 < \ldots < t_{2m-1}$, then m is an even number and $m \geq 2(n + 1)$. In addition, any even value $m \geq 2(n + 1)$ is possible. As a consequence, we have the following Theorem.

Theorem 3.2.

$$\beta_{1,n}^{\text{per}} = \frac{4n^2\pi^2}{T} + \frac{8\pi n(n + 1)}{T} \cot \frac{n\pi}{2(n + 1)} \qquad (3.31)$$

and $\beta_{1,n}^{\text{per}}$ is not attained.

Remark 3.4. We can obtain similar results if we consider $n \in \mathbf{R}^+$ instead of $n \in \mathbf{N}$. Only some minor changes are necessary. From this point of view, if we consider $\beta_{1,n}^{\text{per}}$ as a function of $n \in (0, +\infty)$, then $\lim_{n \to 0+} \beta_{1,n}^{\text{per}} = \frac{16}{T}$, the constant of the classical L^1 Lyapunov inequality at the first eigenvalue which was obtained in [10] by using methods of optimal control theory.

If $n = 0$, the same result can be proved under more general assumptions. In fact, in this case $a \in \Lambda_0$ where

$$\Lambda_0 = \{a \in L_T(\mathbf{R}, \mathbf{R}) \setminus \{0\} : 0 \le \int_0^T a(x)\, dx \text{ and } (3.27) \text{ has nontrivial solutions } \} \tag{3.32}$$

Then $m \ge 2$ and any even value $m \ge 2$ is possible. Consequently

$$\beta_{1,0}^{per} = \inf_{a \in \Lambda_0} \|a^+\|_{L^1(0,T)} = \frac{16}{T}. \tag{3.33}$$

Remark 3.5. The case where $T = 2\pi$ and function a satisfies the condition $A \le a(t) \le B$, a.e. in $(0, 2\pi)$ where $k^2 < A < (k+1)^2 < B$ for some $k \in \mathbf{N} \cup \{0\}$, has been considered in [19], where the authors also use optimal control theory methods. In this paper, the authors define the set $\Lambda_{A,B}$ as the set of functions a such that $A \le a(t) \le B$, a.e. in $(0, T)$ and (3.27) has nontrivial solutions. Then, by using the Pontryagin's maximum principle they prove that the number

$$\beta_{A,B} \equiv \inf_{a \in \Lambda_{A,B}} \|a\|_{L^1(0,T)}$$

is attained. In addition, they calculate $\lim_{B \to +\infty} \beta_{A,B}$.

We can do an analogous study for the anti-periodic boundary value problem

$$u''(t) + a(t)u(t) = 0,\ t \in (0, T),\ u(0) + u(T) = u'(0) + u'(T) = 0, \tag{3.34}$$

where $a \in L_T(\mathbf{R}, \mathbf{R})$. The set of eigenvalues of

$$u''(t) + \lambda u(t) = 0,\ t \in (0, T),\ u(0) + u(T) = u'(0) + u'(T) = 0 \tag{3.35}$$

is given by $\tilde{\lambda}_{2n-1} = \tilde{\lambda}_{2n} = (2n-1)^2 \pi^2 / T^2$, $n \in \mathbf{N}$.

If $n \in \mathbf{N}$ is fixed, we can introduce the set Λ_n^{ant} as

$$\Lambda_n^{ant} = \{a \in L_T(\mathbf{R}, \mathbf{R}) : \tilde{\lambda}_{2n-1} \prec a \text{ and } (3.34) \text{ has nontrivial solutions } \}. \tag{3.36}$$

The similar theorem to Theorem 3.2 is the following one ([6]).

Theorem 3.3.

$$\beta_{1,n}^{ant} \equiv \inf_{a \in \Lambda_n^{ant}} \|a\|_{L^1(0,T)} = \frac{(2n-1)^2 \pi^2}{T} + \frac{2\pi(2n-1)(2n+1)}{T} \cot \frac{(2n-1)\pi}{2(2n+1)} \tag{3.37}$$

and $\beta_{1,n}^{ant}$ is not attained.

Remark 3.6. If $a \in \Lambda_0^{ant}$ where

$$\Lambda_0^{ant} = \{a \in L_T(\mathbf{R}, \mathbf{R}) : \text{(3.34) has nontrivial solutions }\} \tag{3.38}$$

then

$$\beta_{1,0}^{ant} = \inf_{a \in \Lambda_0^{ant}} \|a^+\|_{L^1(0,T)} = \frac{4}{T}. \tag{3.39}$$

Let us remark that the restriction $0 \prec a$ which is natural for the periodic problem (3.27) is not necessary in this case (see Remark 4 in [2]).

3.3   The nth Stability Zone of Linear Periodic Equations

In this section we apply the results of the previous one to the study of stability properties of the Hill's equation

$$u''(t) + a(t)u(t) = 0, \ t \in \mathbf{R}, \tag{3.40}$$

where $a \in L_T(\mathbf{R}, \mathbf{R})$, i.e., a satisfies

$$a : \mathbf{R} \to \mathbf{R} \ \text{is} \ T - \text{periodic and} \ a \in L^1(0, T). \tag{3.41}$$

Remember that the parametric equation

$$u''(t) + (\mu + a(t))u(t) = 0, \ \mu \in \mathbf{R} \tag{3.42}$$

plays a fundamental role in the study of the stability properties of the Hill's equation (3.40) (see Sect. 1.1). Actually, if $\lambda_i(a)$, $i \in \mathbf{N} \cup \{0\}$ and $\tilde{\lambda}_i(a)$, $i \in \mathbf{N}$, denote, respectively, the eigenvalues of

$$u''(t) + (\lambda + a(t))u(t) = 0 \tag{3.43}$$

for the periodic $(u(0) - u(T) = u'(0) - u'(T) = 0)$ and antiperiodic $(u(0) + u(T) = u'(0) + u'(T) = 0)$ problem, then equation (3.42) is stable if

$$\mu \in (\lambda_{2n}(a), \tilde{\lambda}_{2n+1}(a)) \cup (\tilde{\lambda}_{2n+2}(a), \lambda_{2n+1}(a)) \tag{3.44}$$

for some $n \in \mathbf{N} \cup \{0\}$. As a consequence, if either

$$\lambda_{2n}(a) < 0 < \tilde{\lambda}_{2n+1}(a) \tag{3.45}$$

or

$$\tilde{\lambda}_{2n+2}(a) < 0 < \lambda_{2n+1}(a) \tag{3.46}$$

then (3.40) is stable. In this case, we say that $\mu = 0$ belongs to the nth stability zone of (3.42).

Except in very special cases, it is not an easy task to obtain some information on the sign of the previous eigenvalues. It is at this point where the Lyapunov inequalities proved in the previous section can play an important role. In [18] the author shows many other applications of Lyapunov inequalities to eigenvalue problems.

Theorem 3.4.

1. If $a \in L_T(\mathbf{R}, \mathbf{R})$ satisfies

$$\lambda_{2n-1} \prec a, \quad \|a\|_{L^1(0,T)} \leq \beta_{1,n}^{per} \tag{3.47}$$

then

$$\lambda_{2n}(a) < 0 < \lambda_{2n+1}(a) \tag{3.48}$$

2. If $a \in L_T(\mathbf{R}, \mathbf{R})$ satisfies

$$\tilde{\lambda}_{2n-1} \prec a, \quad \|a\|_{L^1(0,T)} \leq \beta_{1,n}^{ant} \tag{3.49}$$

then

$$\tilde{\lambda}_{2n}(a) < 0 < \tilde{\lambda}_{2n+1}(a). \tag{3.50}$$

Proof. For the first part of the theorem, let us assume that the function a satisfies (3.47). Then, since $\lambda_{2n-1} \prec a$, using the variational characterization of the periodic eigenvalues ([18]) we trivially have $\lambda_{2n}(a) < \lambda_{2n}(\lambda_{2n-1}) = 0$.

To prove that $\lambda_{2n+1}(a) > 0$ we use a continuation method: let us define the continuous function $g : [0, 1] \to \mathbf{R}$ by

$$g(\varepsilon) = \lambda_{2n+1}(a_\varepsilon(\cdot)),$$

where $a_\varepsilon(x) = \lambda_{2n-1} + \varepsilon(a(\cdot) - \lambda_{2n-1})$. Then $g(0) = \lambda_{2n+1}(\lambda_{2n-1}) = \lambda_{2n+1} - \lambda_{2n-1} > 0$. Moreover, $g(\varepsilon) \neq 0, \forall \varepsilon \in (0, 1]$. In fact, for each $\varepsilon \in (0, 1]$ the function $a_\varepsilon(x)$ satisfies $\lambda_{2n-1} \prec a_\varepsilon$ and $\|a_\varepsilon(\cdot)\|_{L^1(0,T)} \leq \beta_{1,n}^{per}$. Consequently, we deduce from Theorem 3.1 that the number 0 is not an eigenvalue of the function a_ε for the periodic boundary conditions. As a consequence, $\lambda_{2n+1}(a) = g(1) > 0$.

The second part can be proved in a similar form.

Now, we can exploit the results obtained in the previous Theorem to prove some results on the stability properties of (3.40), of the type $L_\infty - L_1$. The $L_\infty - L_\infty$ results do not permit to the function a to cross the numbers $\frac{p^2\pi^2}{T^2}, p \in \mathbf{N}$ ([1]), but the hypotheses of the next theorem allow to the function a to cross an arbitrary number of them, as long as its L_1-norm is properly controlled (see [6] for a detailed proof).

Theorem 3.5. *Let* $a \in L_T(\mathbf{R}, \mathbf{R})$ *satisfying*

$$\exists p \in \mathbf{N}, \ \exists k \in \left[\frac{p^2\pi^2}{T^2}, \frac{(p+1)^2\pi^2}{T^2} \right] :$$

(3.51)

$$k \leq a, \ \|a\|_{L^1(0,T)} \leq kT + k^{1/2} 2(p+1) \cot \frac{k^{1/2}T}{2(p+1)}.$$

Then $\mu = 0$ *is in the nth stability zone of the Hill's equation (3.42).*

Remark 3.7. The case where $a(t) = \alpha + \beta\psi(t)$, with $\psi \in L_T(\mathbf{R}, \mathbf{R})$, $\int_0^T \psi(t)\, dt = 0$ and $\int_0^T |\psi(t)|\, dt = 1/T$, was studied by Borg ([1]). Borg used the characteristic multipliers determined from Floquet's theory. He deduced stability criteria for (3.42) by using the two parameters α and β. For a concrete function a, this implies the use of the two quantities

$$\frac{1}{T} \int_0^T a(t)\, dt, \quad \frac{1}{T} \left\| a(\cdot) - \frac{1}{T} \int_0^T a(t)\, dt \right\|_{L^1(0,T)}.$$

It is clear that the results given in Theorem 3.5 are of a different nature (see [13] and the translator's note in [11]). In fact, the results of Theorem 3.5 are similar to those obtained by Krein [11] by using a different procedure. However, Krein assumed $k = \frac{p^2\pi^2}{T^2}$ and a strict inequality for $\|a\|_{L^1(0,T)}$ in (3.51) (see Theorem 9 in [11]). It can be assumed a non-strict inequality in (3.51) since the constant $\tilde{\beta}_{1,n}^{\text{ant}}$ is not attained.

Finally, if for a given function $a \in L_T(\mathbf{R}, \mathbf{R})$ we know that a satisfies (3.51), the result given in Theorem 3.5 is more precise than Krein's result since the function

$$kT + k^{1/2} 2(p+1) \cot \frac{k^{1/2}T}{2(p+1)}, \ k \in \left[\frac{p^2\pi^2}{T^2}, \frac{(p+1)^2\pi^2}{T^2} \right]$$

is strictly increasing.

The interested reader can see other stability criteria, based on Lyapunov inequalities, in [6].

3.4 Nonlinear Problems

As we showed in Sect. 2.2, the results on linear equations that we have obtained in Sects. 3.1 and 3.2 can be used to obtain new theorems on the existence and uniqueness of solutions of the corresponding nonlinear boundary value problems.

For example, in the case of Neumann boundary conditions

$$u''(x) + f(x, u(x)) = 0, \ x \in (0, L), \ u'(0) = u'(L) = 0 \tag{3.52}$$

we have the following theorem related to Theorem 2.1 in [14]. This last theorem allows to consider more general boundary value problems, but for ordinary problems with Neumann boundary conditions, our hypotheses allow a more general behavior on the derivative $f_u(x, u)$. We omit the details of the proof since the same ideas and methods of Theorem 2.4 can be used, especially those related to uniqueness of solutions for linear problems (a consequence of Lyapunov inequalities) and the use in the proof of Schauder fixed point theorem.

Theorem 3.6. *Let us consider (3.52) where the following requirements are supposed:*

1. f and f_u are Caratheodory functions on $[0, L] \times \mathbf{R}$ and $f(\cdot, 0) \in L^1(0, L)$
2. There exist functions $\alpha, \beta \in L^\infty(0, L)$, satisfying

$$\lambda_n \leq \alpha(x) \leq f_u(x, u) \leq \beta(x)$$

on $[0, L] \times \mathbf{R}$. Furthermore, α differs from $\lambda_n = \frac{n^2 \pi^2}{L^2}$ on a set of positive measure and β satisfies $\|\beta\|_{L^1(0,L)}\| \leq \beta_{1,n}$, where $\beta_{1,n}$ is given in Theorem 3.1.

Then, problem (3.52) has a unique solution.

We can also consider, for example, nonlinear periodic boundary value problems of the type

$$u''(t) + f(t, u(t)) = 0, \ t \in (0, T), \ u(0) - u(T) = u'(0) - u'(T) = 0. \tag{3.53}$$

and using the same ideas, the following theorem can be proved.

Theorem 3.7. *Let us consider (3.53) where:*

1. f and f_u are Caratheodory function on $\mathbf{R} \times \mathbf{R}$ and $f(t+T, u) = f(t, u), \ \forall \ (t, u) \in \mathbf{R} \times \mathbf{R}$.
2. There exist functions $\alpha, \ \beta \in L^\infty(0, T)$ satisfying

$$\lambda_{2n-1} = \frac{(2n)^2 \pi^2}{T^2} \prec \alpha(t) \leq f_u(t, u) \leq \beta(t), \ \|\beta\|_{L^1(0,T)} \leq \beta_{1,n}^{per} \tag{3.54}$$

where $\beta_{1,n}^{per}$ is given in Theorem 3.2. Then, problem (3.53) has a unique solution.

Clearly, other boundary conditions can be considered (for instance, antiperiodic, mixed ones, etc.). In fact, once we have obtained the Lyapunov constants for linear problems, the procedure is standard: it combines the results obtained in the linear problems with the Schauder fixed point theorem.

Remark 3.8. By using an example in ([12]), it may be seen that the restriction

$$\lambda_{2n-1} \prec \alpha(t) \leq f_u(t, u)$$

in (3.54) cannot be replaced (in nonlinear problems) by the weaker condition

$$\lambda_{2n-1} < f_u(t,u) \le \beta(t).$$

This shows a deep difference (one more) between linear and nonlinear problems.

3.5   Disfocality at Higher Eigenvalues

Under the restriction

$$a \in L^1(0,L), \ \lambda_n \prec a, \tag{3.55}$$

considered in the previous sections, the relation between Neumann boundary conditions and disfocality arises in a natural way. In fact, if $u \in H^1(0,L)$ is any nontrivial solution of

$$u''(x) + a(x)u(x) = 0, \ x \in (0,L), \ u'(0) = u'(L) = 0 \tag{3.56}$$

and the zeros of u are denoted by $x_1 < x_3 < \ldots < x_{2m-1}$, and the zeros of u' are denoted by $0 = x_0 < x_2 < \ldots < x_{2m} = L$, then for each given i, $0 \le i \le 2m - 1$, function u satisfies

$$u''(x)+a(x)u(x) = 0, x \in (x_i,x_{i+1}), \ u(x_i) = 0, \ u'(x_{i+1}) = 0, \ \text{if } i \text{ is odd} \tag{3.57}$$

and

$$u''(x) + a(x)u(x) = 0, \ x \in (x_i,x_{i+1}), \ u'(x_i) = 0, \ u(x_{i+1}) = 0, \ \text{if } i \text{ is even.} \tag{3.58}$$

In consequence, each one of the problems (3.57) and (3.58) with $0 \le i \le 2m - 1$ has nontrivial solution. This simple observation can be used to deduce the following conclusion: if a is any function satisfying (3.55) such that for any $m \ge n + 1$ and any distribution of numbers $0 = x_0 < x_1 < x_2 < \ldots < x_{2m-1} < x_{2m} = L$, either some problem of the type (3.57) or some problem of the type (3.58) has only the trivial solution, then problem (3.56) has only the trivial solution. Lastly, it has been established in [4] (Theorem 2.1 for the case $p = \infty$) that if $b \in L^\infty(c,d)$ satisfies

$$\|b\|_{L^\infty(c,d)} \le \frac{\pi^2}{4(d-c)^2} \ \text{and} \ b \ne \frac{\pi^2}{4(d-c)^2} \ \text{in } (c,d) \tag{3.59}$$

then the unique solution of the boundary value problems

$$u''(x) + b(x)u(x) = 0, \ x \in (c,d), \ u'(c) = u(d) = 0 \tag{3.60}$$

and

$$u''(x) + b(x)u(x) = 0, \ x \in (c,d), \ u(c) = u'(d) = 0 \tag{3.61}$$

is the trivial one.

We may use previous reflections to obtain the following result

Theorem 3.8. *If function a fulfills*

$$a \in L^\infty(0,L), \ \lambda_n \prec a \ and \ \exists \ 0 = y_0 < y_1 < \ldots < y_{2n+1} < y_{2n+2} = L:$$

$$\max \ _{0 \le i \le 2n+1}\{(y_{i+1} - y_i)^2 \|a\|_{L^\infty(y_i,y_{i+1})}\} \le \pi^2/4 \tag{3.62}$$

and, in addition, a is not the constant $\pi^2/4(y_{i+1}-y_i)^2$, at least in one of the intervals $[y_i,y_{i+1}]$, $0 \le i \le 2n+1$,

then the boundary value problem (3.56) has only the trivial solution.

Proof. Take into account that $m \ge n + 1$ and $0 = x_0 < x_1 < x_2 < \ldots < x_{2m-1} < x_{2m} = L$ is any arbitrary distribution of numbers, then or

$$[x_j, x_{j+1}] \subset [y_i, y_{i+1}], \ \text{strictly} \tag{3.63}$$

for some $0 \le i \le 2n + 1$, $0 \le j \le 2m - 1$ or

$$m = n + 1 \ and \ x_i = y_i, \ \forall \ 0 \le i \le 2n + 2. \tag{3.64}$$

If (3.63) is satisfied, then

$$\|a\|_{L^\infty(x_j,x_{j+1})} < \|a\|_{L^\infty(y_i,y_{i+1})} \le \frac{\pi^2}{4(y_{i+1} - y_i)^2} < \frac{\pi^2}{4(x_{j+1} - x_j)^2} \tag{3.65}$$

and consequently we deduce from (3.57), (3.58) and (3.59) that (3.56) has only the trivial solution.

If (3.64) is satisfied, we deduce from the hypotheses of the theorem that a is not the constant $\pi^2/4(x_{i+1} - x_i)^2$, at least in one of the intervals $[x_i, x_{i+1}]$, $0 \le i \le 2n + 1$. Therefore, again (3.57), (3.58), and (3.59) imply that (3.56) has only the trivial solution. In any case, we have the desired conclusion.

Remark 3.9. If in previous theorem we choose $y_i = \frac{iL}{2(n+1)}$, $0 \le i \le 2n + 2$, then we have the so-called nonuniform nonresonance conditions at higher eigenvalues [7, 14] but if for instance, $y_{j+1} - y_j < \frac{L}{2(n+1)}$, for some j, $0 \le j \le 2n + 1$, function a can satisfy $\|a\|_{L^\infty(y_j,y_{j+1})} = \frac{\pi^2}{4(y_{j+1}-y_j)^2}$ (which is a quantity greater than $\lambda_{n+1} = \frac{(n+1)^2\pi^2}{L^2}$) as long as a satisfies (3.62) for each $i \neq j$.

Remark 3.10. The hypothesis of the previous theorem is optimal in the sense that if a is the constant $\pi^2/4(y_{i+1} - y_i)^2$ in each one of the intervals (y_i, y_{i+1}), $0 \le i \le 2n + 1$, then (3.56) has nontrivial solutions (see [5], Remark 7).

To conclude, we comment some relations between the Lyapunov constant $\beta_{1,n}$, given in Theorem 3.1 and disfocality. To this respect, it is clear from the definition of $\beta_{1,n}$ that if a function a satisfies

$$a \in L^1(0,L), \ \lambda_n \prec a, \ \|a - \lambda_n\|_1 < \beta_{1,n} \tag{3.66}$$

then the unique solution of (3.56) is the trivial one. In the next theorem we prove that, with the use of disfocality, we can obtain a more general condition.

Theorem 3.9. *1. If function $a \in L^1(0,L)$, $\lambda_n \prec a$, satisfies:*

$$\exists \ 0 = y_0 < y_1 < \ldots < y_{2n+1} < y_{2n+2} = L :$$

$$y_{i+1} - y_i < \tfrac{L}{2n} ; \ \|a - \lambda_n\|_{L^1(y_i,y_{i+1})} < \tfrac{n\pi}{L} \cot \tfrac{n\pi(y_{i+1}-y_i)}{L}, \ \forall \ 0 \le i \le 2n + 1 \tag{3.67}$$

then the unique solution of (3.56) is the trivial one.
2. (3.66) implies (3.67).
3. If $0 = y_0 < y_1 < \ldots < y_{2n+1} < y_{2n+2} = L$, is any distribution of numbers such that $y_{k+1} - y_k < \tfrac{L}{2n}$, $\forall \ 0 \le k \le 2n + 1$ and $y_{i+1} - y_i \ne y_{j+1} - y_j$, for some $0 \le i,j \le 2n + 1$, then there exists $a \in L^1(0,L)$, $\lambda_n \prec a$, satisfying (3.67) but not satisfying (3.66).

Proof. If a satisfies (3.67), then the unique solution of (3.56) is the trivial one. In fact, if this is not true, let u be a nontrivial solution of (3.56) and let us denote the zeros of u by $x_1 < x_3 < \ldots < x_{2m-1}$ and the zeros of u' by $0 = x_0 < x_2 < \ldots < x_{2m} = L$. Since $m \ge n + 1$, then

$$[x_j, x_{j+1}] \subset [y_i, y_{i+1}] \tag{3.68}$$

for some $0 \le i \le 2n + 1$, $0 \le j \le 2m - 1$. Consequently,

$$\frac{\|a - \lambda_n\|_{L^1(x_j,x_{j+1})}}{\cot \frac{n\pi(x_{j+1}-x_j)}{L}} \le \frac{\|a - \lambda_n\|_{L^1(y_i,y_{i+1})}}{\cot \frac{n\pi(y_{i+1}-y_i)}{L}} < \frac{n\pi}{L}.$$

From here we deduce

$$\|a - \lambda_n\|_{L^1(x_j,x_{j+1})} < \frac{n\pi}{L} \cot \frac{n\pi(x_{j+1}-x_j)}{L}$$

which is a contradiction with Lemmas 3.1 and 3.2.

Next we prove that (3.66) implies (3.67). We can certainly assume that $\inf a > \lambda_n$, for if not, we replace a by $a + \delta$ (for small $\delta > 0$) and the new function $a + \delta$ satisfies (3.66). Note that if condition (3.67) is satisfied for $a + \delta$ then it is also satisfied for the function a.

Now choose $\varepsilon > 0$ sufficiently small. Since the function

$$\frac{\|a - \lambda_n\|_{L^1(0,y)}}{\cot \frac{n\pi(y-0)}{L}}$$

is strictly increasing with respect to $y \in (0, \frac{L}{2n})$ and

$$\lim_{y \to 0^+} \frac{\|a - \lambda_n\|_{L^1(0,y)}}{\cot \frac{n\pi(y-0)}{L}} = 0, \quad \lim_{y \to \frac{L}{2n}^-} \frac{\|a - \lambda_n\|_{L^1(0,y)}}{\cot \frac{n\pi(y-0)}{L}} = +\infty$$

there is a unique y_1, $0 = y_0 < y_1 < \frac{L}{2n}$ such that

$$\frac{\|a - \lambda_n\|_{L^1(0,y_1)}}{\cot \frac{n\pi(y_1-0)}{L}} = \frac{n\pi}{L} - \varepsilon. \tag{3.69}$$

With the help of a similar reasoning, it is possible to prove the existence of points $0 = y_0 < y_1 < \ldots < y_{2n+1}$, such that

$$\frac{\|a - \lambda_n\|_{L^1(y_i,y_{i+1})}}{\cot \frac{n\pi(y_{i+1}-y_i)}{L}} = \frac{n\pi}{L} - \varepsilon, \quad y_{i+1} - y_i < \frac{L}{2n}, \quad 0 \le i \le 2n. \tag{3.70}$$

(If it is necessary, we can define $a(x) = \lambda_n$, $\forall x > L$).

Since $y_{i+1} - y_i < \frac{L}{2n}$, $0 \le i \le 2n - 1$, then $y_{2n} < L$.

If $y_{2n+1} \ge L$, then we replace the number y_{2n+1} with $y_{2n+1} = L - \mu$ (for small $\mu > 0$). Finally, choosing $y_{2n+2} = L$, we obtain (3.67).

If $y_{2n+1} < L$, take $y_{2n+2} = L$. We claim that

$$y_{2n+2} - y_{2n+1} < \frac{L}{2n} \quad \text{and} \quad \frac{\|a - \lambda_n\|_{L^1(y_{2n+1},y_{2n+2})}}{\cot \frac{n\pi(y_{2n+2}-y_{2n+1})}{L}} < \frac{n\pi}{L} - \varepsilon. \tag{3.71}$$

In fact, if $y_{2n+2} - y_{2n+1} \ge \frac{L}{2n}$, then $y_{2n+1} \le \frac{L(2n-1)}{2n}$. Then, from (3.70), Lemma 3.3 (with $r = 2n+1$, $S = \frac{n\pi}{L}(y_{2n+1})$ and $z_i = \frac{n\pi}{L}(y_{i+1} - y_i)$) and using the monotonicity of cot in $(0, \pi/2)$ we obtain

$$\frac{n\pi}{L} 2(n+1) \cot \frac{n\pi}{2(n+1)} = \beta_{1,n} > \sum_{i=0}^{2n} \|a - \lambda_n\|_{L^1(y_i,y_{i+1})}$$

$$= \left(\frac{n\pi}{L} - \varepsilon\right) \sum_{i=0}^{2n} \cot \frac{n\pi}{L}(y_{i+1} - y_i)$$

$$\ge \left(\frac{n\pi}{L} - \varepsilon\right)(2n+1) \cot \frac{n\pi}{L(2n+1)} y_{2n+1}$$

$$\ge \left(\frac{n\pi}{L} - \varepsilon\right)(2n+1) \cot \frac{\pi(2n-1)}{2(2n+1)}.$$

If $\varepsilon \to 0^+$, we conclude

$$\beta_{1,n} \geq \frac{n\pi}{L}(2n+1)\cot\frac{\pi(2n-1)}{2(2n+1)}. \tag{3.72}$$

Now, by using that the function $x \mapsto \dfrac{2\pi \cot x}{\pi - 2x}$ is strictly decreasing in $(0, \pi/2)$ and that $\dfrac{\pi(2n-1)}{2(2n+1)} < \dfrac{n\pi}{2(n+1)}$, we obtain

$$\beta_{1,n} \geq \frac{n\pi}{L}(2n+1)\cot\frac{\pi(2n-1)}{2(2n+1)} > \frac{n\pi}{L}2(n+1)\cot\frac{n\pi}{2(n+1)} = \beta_{1,n}$$

which is a contradiction.

It remains to prove the second part of the claim (3.71). In fact, if this second part is not true, then from (3.70) and Lemma 3.3 (with $r = 2n+2$, $S = n\pi$ and $z_i = \frac{n\pi}{L}(y_{i+1} - y_i)$) we have

$$\begin{aligned}
\|a - \lambda_n\|_{L^1(0,L)} &= \sum_{i=0}^{2n+1} \|a - \lambda_n\|_{L^1(y_i, y_{i+1})} \\
&\geq \left(\frac{n\pi}{L} - \varepsilon\right)\sum_{i=0}^{2n+1}\cot\frac{n\pi(y_{i+1} - y_i)}{L} \geq \left(\frac{n\pi}{L} - \varepsilon\right)\frac{\beta_{1,n}}{n\pi/L},
\end{aligned}$$

for $\varepsilon > 0$ sufficiently small. This is a contradiction with (3.66).

Finally, to prove part (3) of the theorem, let us take numbers $0 = y_0 < y_1 < \dots < y_{2n+1} < y_{2n+2} = L$, such that $y_{k+1} - y_k < \frac{L}{2n}$, $\forall\, 0 \leq k \leq 2n+1$ and $y_{i+1} - y_i \neq y_{j+1} - y_j$, for some $0 \leq i, j \leq 2n+1$. Then from Lemma 3.3 we obtain

$$\sum_{i=0}^{2n+1}\frac{n\pi}{L}\cot\frac{n\pi(y_{i+1} - y_i)}{L} > \frac{2\pi n(n+1)}{L}\cot\frac{n\pi}{2(n+1)} = \beta_{1,n}.$$

Now, choose a function $a \in L^1(0, L)$, $\lambda_n \prec a$, satisfying

$$\|a - \lambda_n\|_{L^1(y_i, y_{i+1})} = \frac{n\pi}{L}\cot\frac{n\pi(y_{i+1} - y_i)}{L} - \varepsilon, \quad \forall\, 0 \leq i \leq 2n+1$$

It is trivial that if ε is sufficiently small, then function a satisfies (3.67) whereas

$$\|a - \lambda_n\|_{L^1(0,L)} = \sum_{i=0}^{2n+1}\|a - \lambda_n\|_{L^1(y_i, y_{i+1})} > \beta_{1,n}.$$

We finish this chapter by showing how to use previous reasonings to obtain new theorems on the existence and uniqueness of solutions of nonlinear b.v.p.

$$u''(x) + f(x, u(x)) = 0, \ x \in (0, L), \ u'(0) = u'(L) = 0. \tag{3.73}$$

For example, we have the following theorem related to Theorem 2.1 in [14]. This last theorem allows to consider more general boundary value problems, but for ordinary problems with Neumann boundary conditions our hypotheses allow a more general behavior on the derivative $f_u(x, u)$. We omit the details of the proof (see [2, 3] for similar results at the two first eigenvalues).

Theorem 3.10. *Let us consider (3.73) where the following requirements are supposed:*

1. f and f_u are Caratheodory functions on $[0, L] \times \mathbf{R}$ and $f(\cdot, 0) \in L^1(0, L)$.
2. There exist functions $\alpha, \beta \in L^\infty(0, L)$, satisfying

$$\lambda_n \leq \alpha(x) \leq f_u(x, u) \leq \beta(x)$$

on $[0, L] \times \mathbf{R}$. Furthermore, α differs from λ_n on a set of positive measure and β satisfies either hypothesis (3.62) of Theorem 3.8 or hypothesis (3.67) of Theorem 3.9.

Then, problem (3.73) has a unique solution.

References

1. Borg, G.: Über die Stabilität gewisser Klassen von linearen Differentialgleichungen. Ark. Mat. Astr. Fys. **31 A**(1), 1–31 (1944)
2. Cañada, A., Montero, J.A., Villegas, S.: Lyapunov type inequalities and Neumann boundary value problems at resonance. Math. Inequal. Appl. **8**, 459–475 (2005)
3. Cañada, A., Montero, J.A., Villegas, S.: Lyapunov inequalities for partial differential equations. J. Funct. Anal. **237**, 176–193 (2006)
4. Cañada, A., Villegas, S.: Optimal Lyapunov inequalities for disfocality and Neumann boundary conditions using L^p norms. Discrete Contin. Dyn. Syst. Ser. A **20**, 877–888 (2008)
5. Cañada, A., Villegas, S.: Lyapunov inequalities for Neumann boundary conditions at higher eigenvalues. J. Eur. Math. Soc. **12**, 163–178 (2010).
6. Cañada, A., Villegas, S.: Lyapunov inequalities for the periodic boundary value problem at higher eigenvalues. J. Math. Anal. Appl. **376**, 429–442 (2011)
7. Dolph, C.L.: Nonlinear integral equations of Hammerstein type. Trans. Am. Math. Soc. **66**, 289–307 (1949)
8. Hartman, P.: Ordinary Differential Equations. Wiley, New York (1964)
9. Huaizhong, W., Yong, L.: Two point boundary value problems for second order ordinary differential equations across many resonance points. J. Math. Anal. Appl. **179**, 61–75 (1993)
10. Huaizhong, W., Yong, L.: Existence and uniqueness of periodic solutions for Duffing equations across many points of resonance. J. Differ. Equ. **108**, 152–169 (1994)

11. Krein, M.G.: On certain problems on the maximum and minimum of characteristic values and on the Lyapunov zones of stability. American Mathematical Society Translations, Series 2, vol. 1. American Mathematical Society, Providence, RI (1955)
12. Lazer, A.C., Leach, D.E.: On a nonlinear two-point boundary value problem. J. Math. Anal. Appl. **26**, 20–27 (1969)
13. Magnus, W., Winkler, S.: Hill's Equation. Dover, New York (1979)
14. Mawhin, J., Ward, J.R.: Nonresonance and existence for nonlinear elliptic boundary value problems. Nonlinear Anal. **5**, 677–684 (1981)
15. Mawhin, J., Ward, J.R.: Nonuniform nonresonance conditions at the two first eigenvalues for periodic solutions of forced Liénard and Duffing equations. Rocky Mountain J. Math. **12**, 643–654 (1982)
16. Mawhin, J., Ward, J.R.: Periodic solutions of some forced Liénard differential equations at resonance. Arch. Math. (Basel) **41**, 337–351 (1983)
17. Mawhin, J., Ward, J.R., Willem, M.: Variational methods and semilinear elliptic equations. Arch. Ration. Mech. Anal. **95**, 269–277 (1986)
18. Pinasco, J.P.: Lyapunov-Type Inequalities. With Applications to Eigenvalue Problems. Springer Briefs in Mathematics. Springer, New York (2013)
19. Wang, H., Li, Y.: Periodic solutions for Duffing equations. Nonlinear Anal. **24**, 961–979 (1995)
20. Yong, L., Huaizhong, W.: Neumann problems for second order ordinary differential equations across resonance. Z. Angew Math. Phys. **46**, 393–406 (1995)
21. Yuhua, L., Yong, L., Qinde, Z.: Second boundary value problems for nonlinear ordinary differential equations across resonance. Nonlinear Anal. **28**, 999–1009 (1997)

Chapter 4
Partial Differential Equations

Abstract This chapter is devoted to the study of L_p Lyapunov-type inequalities ($1 \le p \le +\infty$) for linear partial differential equations. More precisely, we treat the case of Neumann boundary conditions on bounded and regular domains in $\mathbf{R}^N$. In the case of Dirichlet conditions, it is possible to obtain analogous results in an easier way. We also treat the case of higher eigenvalues in the radial case, by using appropriate minimizing sequences and a detailed analysis about the number and distribution of zeros of radial nontrivial solutions. It is proved that the relation between the quantities p and $N/2$ plays a crucial role just to have nontrivial Lyapunov inequalities. This fact shows a deep difference with respect to the ordinary case. The linear study is combined with Schauder fixed point theorem to provide new conditions about the existence and uniqueness of solutions for resonant nonlinear problems.

4.1 L_1-Lyapunov Inequality is Meaningless for PDEs

In the previous chapters we have considered ordinary differential equations and obtained different Lyapunov inequalities. A natural question is whether it is possible to obtain similar results for partial differential equations. For instance, there does exist an L_1-Lyapunov inequality for the P.D.E version of Eq. (2.1)? In this chapter we will see that the answer to this question is negative by constructing explicit examples of nontrivial solutions of linear equations (see Theorem 4.1 below). That is, the L_1 norm of the linear coefficient $a(x)$ could be arbitrary small, contrary to the one-dimensional case.

This fact inspires us the following question: for which values $1 \le p \le \infty$ does exist an L_p-Lyapunov inequality for the N- dimensional case? Through this chapter we will see that the answer to this question only depends on the relation between p and $N/2$, but not on the considered domain $\Omega \subset \mathbf{R}^N$ [2].

© The Author(s) 2015

A. Cañada, S. Villegas, *A Variational Approach to Lyapunov Type Inequalities*,
SpringerBriefs in Mathematics, DOI 10.1007/978-3-319-25289-6_4

4.2 L_p-Lyapunov Inequalities

Consider the linear problem

$$\left.\begin{array}{l} -\Delta u(x) = a(x)u(x),\ x \in \Omega \\ \frac{\partial u}{\partial n}(x) = 0, \qquad x \in \partial\Omega \end{array}\right\},\tag{4.1}$$

where $\Omega \subset \mathbf{R}^N$ ($N \geq 2$) is a bounded and regular domain and the function $a : \Omega \to \mathbf{R}$ belongs to the set Λ defined as

$$\Lambda = \{a \in L^{\frac{N}{2}}(\Omega) \setminus \{0\} : \int_\Omega a(x)\ dx \geq 0 \text{ and } (4.1) \text{ has nontrivial solutions}\}\tag{4.2}$$

if $N \geq 3$ and

$$\Lambda = \{a : \Omega \to \mathbf{R} \text{ s. t. } \exists\, q \in (1, \infty] \text{ with } a \in L^q(\Omega) \setminus \{0\}, \int_\Omega a(x)\ dx \geq 0 \text{ and } (4.1)$$

has nontrivial solutions}
if $N = 2$.

We will consider weak solutions of (4.1). Remind that we say that $u \in H^1(\Omega)$ is a weak solution of (4.1) if

$$\int_\Omega \nabla u \cdot \nabla v = \int_\Omega a\, u\, v, \ \forall\, v \in H^1(\Omega).\tag{4.3}$$

This definition has sense since we have imposed in the definition of Λ that $a \in L^{N/2}$ if $N \geq 3$ and $a \in L^q$ for some $q > 1$, if $N = 2$. In effect, by the standard Sobolev embedding $H^1(\Omega) \subset L^{2N/(N-2)}(\Omega)$, for $N \geq 3$, it is guaranteed that $a(x)u(x)v(x)$ belongs to $L^1(\Omega)$, for every $u, v \in H^1(\Omega)$ and $a \in L^{N/2}(\Omega)$ (similar reasoning for $N = 2$).

On the other hand, it is also natural to suppose $\int_\Omega a(x)\ dx \geq 0$ in the definition of Λ, as the one-dimensional case (see Remark 2.1). To see this, consider any nonconstant positive function $u_0 \in C^2(\overline{\Omega})$ satisfying $\frac{\partial u_0}{\partial n}(x) = 0$, $x \in \partial\Omega$. Define $u_n = u_0 + n$, $a_n = (-\Delta u_n)/u_n$, for every $n \in \mathbf{N}$. It is clear that, for every natural number n, u_n is a solution of (4.1) for $a = a_n$. Integrating by parts, we obtain that $\int_\Omega a_n = \int_\Omega (-\Delta u_n)/u_n = -\int_\Omega |\nabla u_n|^2/u_n^2 < 0$. Observe that, for every $1 \leq p \leq \infty$ we have that $\|a_n\|_p = \|(-\Delta u_0)/(u_0 + n)\|_p$ tends to 0, as n tends to ∞. Hence, if we expect to have some nontrivial Lyapunov inequalities, we should avoid this class of functions with negative integral.

Since the positive eigenvalues of the eigenvalue problem

$$\left.\begin{array}{l} -\Delta u(x) = \lambda u(x),\ x \in \Omega \\ \frac{\partial u}{\partial n}(x) = 0, \qquad x \in \partial\Omega \end{array}\right\}\tag{4.4}$$

belong to Λ, the quantity

$$\beta_p \equiv \inf_{a \in \Lambda \cap L^p(\Omega)} \|a^+\|_p, \ 1 \le p \le \infty \tag{4.5}$$

is well defined and it is a nonnegative real number.

The first novelty in partial differential equations that we will observe is that $\beta_1 = 0$ for each $N \ge 2$. Moreover, we prove that if $N = 2$, then $\beta_p > 0$, $\forall\, p \in (1, \infty]$ and that if $N \ge 3$, then $\beta_p > 0$ if and only if $p \ge N/2$. Also, for each $N \ge 2$, β_p is attained if $p > N/2$. These results show a great difference with respect to the ordinary case, where $\beta_p > 0$ for each $1 \le p \le \infty$. Moreover, we prove some qualitative properties of β_p such as the continuity and monotonicity with respect to p.

Theorem 4.1. *Let $N \ge 2$, $1 \le p \le \infty$. The following statements hold:*

1. *If $N = 2$ then $\beta_p > 0 \Leftrightarrow 1 < p \le \infty$*
   If $N \ge 3$ then $\beta_p > 0 \Leftrightarrow \frac{N}{2} \le p \le \infty$
2. *If $\frac{N}{2} < p \le \infty$, then β_p is attained. In this case, any function $a \in \Lambda \cap L^p(\Omega)$ in which β_p is attained is of the form*

   i) *$a(x) \equiv \lambda_1$, if $p = \infty$, where λ_1 is the first strictly positive eigenvalue of (4.4).*
   ii) *$a(x) \equiv |u(x)|^{\frac{2}{p-1}}$, if $\frac{N}{2} < p < \infty$, where u is a solution of the problem*

$$\left. \begin{array}{ll} -\Delta u(x) = |u(x)|^{\frac{2}{p-1}} u(x), & x \in \Omega \\ \frac{\partial u}{\partial n}(x) = 0, & x \in \partial\Omega \end{array} \right\} \tag{4.6}$$

3. *The mapping $(\frac{N}{2}, \infty) \to \mathbf{R}$, $p \mapsto \beta_p$, is continuous and the mapping $[\frac{N}{2}, \infty) \to \mathbf{R}$, $p \mapsto |\Omega|^{-1/p}\beta_p$, is strictly increasing.*
4. *There exists always the limits $\lim_{p \to \infty} \beta_p$ and $\lim_{p \to \frac{N}{2}^+} \beta_p$ and take the values*

   i) *$\lim_{p \to \infty} \beta_p = \beta_\infty$, if $N \ge 2$*
   ii) *$\lim_{p \to \frac{N}{2}^+} \beta_p \ge \beta_{\frac{N}{2}} > 0$, if $N \ge 3$*
   $\lim_{p \to 1+} \beta_p = 0$, if $N = 2$

Remark 4.1. Note that the only left open question of the previous theorem is the attainability of $\beta_{N/2}$, for $N \ge 3$. Up to our knowledge this is still not solved. On the other hand, as we have mentioned before, all the results above remain valid if we consider Dirichlet boundary conditions. In fact, the proof of the previous theorem is simpler than in the case of Neumann boundary conditions that we have considered. However, for Dirichlet boundary conditions, it is proved [9] that, for $N \ge 3$, $\beta_{N/2}$ is not attained. In fact, an explicit calculus of this value is given: $\beta_{N/2} = S_N$, where S_N is the best Sobolev constant in $\mathbf{R}^N$:

$$S_N = \pi N(N - 2) \left[\frac{\Gamma(N/2)}{\Gamma(N)} \right]^{2/N}.$$

Remark 4.2. It seems difficult to obtain explicit expressions for β_p, as a function of p, Ω and N, even for simple cases of domains, like balls or cubes. This is also a difference with respect to the one-dimensional case, in which explicit values of β_p were obtained (see Theorem 2.3).

For the proof of parts 1 and 2 of Theorem 4.1, we will distinguish three cases: the subcritical case ($1 \leq p < \frac{N}{2}$ if $N \geq 3$, and $p = 1$ if $N = 2$), the critical case ($p = \frac{N}{2}$ if $N \geq 3$) and the supercritical case ($p > \frac{N}{2}$ if $N \geq 2$).

4.2.1   The Subcritical Case

In this subsection, we study the subcritical case, i.e., $1 \leq p < \frac{N}{2}$, if $N \geq 3$, and $p = 1$ if $N = 2$. In all these cases we will prove that $\beta_p = 0$. Roughly speaking, the main idea of the proof is to take first a nontrivial function u and to calculate the corresponding function a for which u is a solution of (4.1). Obviously, if u is smooth enough, then we must impose two conditions: (i) $\frac{\partial u}{\partial n} = 0$ on $\partial \Omega$, and (ii) the zeros of u are also zeros of Δu. For instance, if $\Omega = B(0, 1)$ (the open ball in $\mathbf{R}^N$ of center zero and radius one) we can take radial functions $u(x) = f(|x|)$ of the form $f(r) = Ar^{-a} + Br^{-b}$, ($r \in (\varepsilon, 1]$) for certain a, b, A, B such that the two mentioned conditions are satisfied.

Lemma 4.1. *Let $N \geq 3$ and $1 \leq p < \frac{N}{2}$. Then $\beta_p = 0$*

Proof. First of all, note that if we define $\Omega + x_0 = \{x + x_0 \, ; \, x \in \Omega\}$ (for arbitrary $x_0 \in \mathbf{R}^N$), then $\beta_p (\Omega + x_0) = \beta_p (\Omega)$. On the other hand, if we define $r\Omega = \{rx \, ; \, x \in \Omega\}$ (for arbitrary $r \in \mathbf{R}^+$), then $\beta_p (r\Omega) = r^{\frac{N}{p} - 2} \beta_p (\Omega)$. Hence $\beta_p(\Omega) = 0 \Leftrightarrow \beta_p (r\Omega + x_0) = 0$. Then, we can suppose without loss of generality that $\overline{B}(0, 1) \subset \Omega$.

Take now arbitrary real numbers $a > b > 0$ satisfying $a + b = N - 2$ and choose $0 < \varepsilon < \left(\frac{a}{b}\right)^{\frac{1}{b-a}} < 1$. Define $u : \Omega \to \mathbf{R}$ as the radial function

$$u(x) = \begin{cases} -\alpha|x|^2 + \beta, & \text{if } |x| \leq \varepsilon \\ b|x|^{-a} - a|x|^{-b}, & \text{if } \varepsilon < |x| < 1 \\ b - a, & \text{if } |x| \geq 1, \, x \in \Omega, \end{cases} \tag{4.7}$$

where α and β are defined such that $u \in C^1(\overline{\Omega})$; i.e., $\alpha = \frac{ab}{2}\left(\varepsilon^{-a-2} - \varepsilon^{-b-2}\right) > 0$ and $\beta = \varepsilon^{-a}b\left(\frac{a}{2} + 1\right) - \varepsilon^{-b}a\left(\frac{b}{2} + 1\right) > 0$.

It is straightforward to check that u is a solution of (4.1), being $a : \Omega \to \mathbf{R}$ the radial function

$$a(x) = \begin{cases} \dfrac{2N\alpha}{-\alpha|x|^2 + \beta}, & \text{if } |x| \le \varepsilon \\[3mm] \dfrac{ab}{|x|^2}, & \text{if } \varepsilon < |x| < 1 \\[3mm] 0, & \text{if } |x| \ge 1,\ x \in \Omega. \end{cases} \tag{4.8}$$

It is easily seen that $a(x) \ge 0$ and $a(x) \in L^\infty(\Omega)$. Hence $a(x) \in \Lambda$. Let us estimate the L_p-norm of $a(x)$. To this aim, taking into account that the maximum of $a(x)$ in $\overline{B}(0, \varepsilon)$ is attained in $|x| = \varepsilon$, we have

$$\|a\|_p \le \left(\int_{B(0,\varepsilon)} \left(\frac{2N\alpha}{-\alpha\varepsilon^2 + \beta} \right)^p + \int_{B(0,1)\setminus B(0,\varepsilon)} \left(\frac{ab}{|x|^2} \right)^p \right)^{\frac{1}{p}}$$

$$= \left(\left(\frac{Nab \left(\varepsilon^{-a-2} - \varepsilon^{-b-2} \right)}{b\varepsilon^{-a} - a\varepsilon^{-b}} \right)^p \frac{w_N}{N} \varepsilon^N + \frac{(ab)^p w_N \left(1 - \varepsilon^{N-2p} \right)}{N - 2p} \right)^{\frac{1}{p}}. \tag{4.9}$$

Then β_p is smaller than this expression. But (for fixed real numbers $a > b > 0$ with $a + b = N - 2$) we can take limit when ε tends to zero in (4.9). This gives (taking into account that $p < N/2$):

$$\beta_p \le \frac{ab\, w_N^{\frac{1}{p}}}{(N - 2p)^{\frac{1}{p}}}.$$

Finally, taking limit when b tends to zero in the last formula, we conclude $\beta_p = 0$.

Lemma 4.2. *Let $N = 2$ and $p = 1$. Then $\beta_1 = 0$.*

Proof. As we have argued in Lemma 4.1 it is easy to check that $\beta_1 (r\Omega + x_0) = \beta_1 (\Omega)$, for every $x_0 \in \mathbf{R}^2, r \in \mathbf{R}^+$. Then, we can suppose again without loss of generality that $\overline{B}(0, 1) \subset \Omega$.

Take now arbitrary real numbers $K > \log(4)$ and $\varepsilon > 0$ satisfying $\log(\varepsilon^2) + K < 0$. Define $u : \Omega \to \mathbf{R}$ as the radial function

$$u(x) = \begin{cases} \left(\dfrac{|x|}{\varepsilon} \right)^2 + \log(\varepsilon^2) + K - 1, & \text{if } |x| \le \varepsilon \\[3mm] \log(|x|^2) + K, & \text{if } \varepsilon < |x| \le \dfrac{1}{2} \\[3mm] -4\,(1 - |x|)^2 + 1 + K - \log(4), & \text{if } \dfrac{1}{2} < |x| < 1 \\[3mm] 1 + K - \log(4), & \text{if } |x| \ge 1,\ x \in \Omega. \end{cases} \tag{4.10}$$

Then, it is easy to check that u is a solution of (4.1), being $a : \Omega \to \mathbf{R}$ the radial function

$$a(x) = \begin{cases} \dfrac{-4}{|x|^2 + \varepsilon^2 \left(\log(\varepsilon^2) + K - 1\right)}, & \text{if } |x| \le \varepsilon \\[3mm] 0, & \text{if } \varepsilon < |x| \le \dfrac{1}{2} \\[3mm] \dfrac{16 - \frac{8}{|x|}}{-4 \left(1 - |x|\right)^2 + 1 + K - \log(4)}, & \text{if } \dfrac{1}{2} < |x| < 1 \\[3mm] 0, & \text{if } |x| \ge 1, \ x \in \Omega. \end{cases} \qquad (4.11)$$

Trivially, $a(x) \ge 0$ and $a(x) \in L^\infty(\Omega)$. Hence $a(x) \in \Lambda$. Let us estimate the L_1-norm of $a(x)$:

$$\|a\|_1 = \int_{B(0,\varepsilon)} a(x)\,dx + \int_{B(0,1) \backslash B(0,\frac{1}{2})} a(x)\,dx$$

$$= 2\pi \int_0^\varepsilon \frac{-4r\,dr}{r^2 + \varepsilon^2 \left(\log(\varepsilon^2) + K - 1\right)} + 2\pi \int_{\frac{1}{2}}^1 \frac{(16r - 8)\,dr}{-4\left(1 - r\right)^2 + 1 + K - \log(4)}.$$

It is possible to evaluate the first integral and to estimate the second one:

$$\|a\|_1 \le 4\pi \log\left(\frac{\varepsilon^2 \left(\log(\varepsilon^2) + K - 1\right)}{\varepsilon^2 + \varepsilon^2 \left(\log(\varepsilon^2) + K - 1\right)}\right) + 2\pi \int_{\frac{1}{2}}^1 \frac{(16 \cdot 1 - 8)\,dr}{-4\left(1 - \frac{1}{2}\right)^2 + 1 + K - \log(4)}.$$

Then β_1 is smaller than this expression. But (for fixed real number $K > \log(4)$) we can take limit when ε tends to zero in this formula. This gives

$$\beta_1 \le \frac{8\pi}{K - \log(4)}.$$

Finally, taking limit when K tends to $+\infty$ we conclude $\beta_1 = 0$.

4.2.2   The Critical Case

In this subsection we study the critical case, i.e., $p = \frac{N}{2}$, if $N \ge 3$. We will prove that $\beta_{N/2} > 0$.

Lemma 4.3. *If $N \ge 3$ and $p = \frac{N}{2}$ we have $\beta_p > 0$.*

Proof. If $a \in \Lambda$ and $u \in H^1(\Omega)$ is a nontrivial solution of (4.1), then taking respectively $v = u$ and $v=1$ in (4.3) we obtain

$$\int_\Omega |\nabla u|^2 = \int_\Omega au^2, \quad \int_\Omega a\,u = 0. \tag{4.12}$$

Therefore, since $\int_\Omega a \geq 0$ it is deduced for each $k \in \mathbf{R}$ that

$$\int_\Omega |\nabla(u+k)|^2 = \int_\Omega |\nabla u|^2 = \int_\Omega au^2 \leq \int_\Omega au^2 + k^2 \int_\Omega a$$

$$= \int_\Omega a(u+k)^2 \leq \int_\Omega a^+(u+k)^2. \tag{4.13}$$

It follows from Hölder inequality

$$\int_\Omega |\nabla(u+k)|^2 \leq \|a^+\|_{\frac{N}{2}} \|(u+k)^2\|_{\frac{N}{N-2}}.$$

Also, since u is a nonconstant solution of (4.1), $u + k$ is a nontrivial function. Consequently

$$\|a^+\|_{\frac{N}{2}} \geq \frac{\displaystyle\int_\Omega |\nabla(u+k)|^2}{\|u+k\|^2_{\frac{2N}{N-2}}}.$$

Now, choose $k_0 \in \mathbf{R}$ satisfying $\int_\Omega (u+k_0) = 0$. Then,

$$\|a^+\|_{\frac{N}{2}} \geq \frac{\displaystyle\int_\Omega |\nabla(u+k_0)|^2}{\|u+k_0\|^2_{\frac{2N}{N-2}}} \geq \inf_{v \in X_\infty \setminus \{0\}} \frac{\displaystyle\int_\Omega |\nabla v|^2}{\|v\|^2_{\frac{2N}{N-2}}} = C, \ \forall\, a \in \Lambda,$$

where $X_\infty = \{v \in H^1(\Omega) : \int_\Omega v = 0\}$.

Finally, the continuous inclusions $X_\infty \subset H^1(\Omega) \subset L^{\frac{2N}{N-2}}(\Omega)$ gives us $C > 0$, which completes the proof.

4.2.3 The Supercritical Case

In this subsection, we study the supercritical case, i.e., $p > \frac{N}{2}$, if $N \geq 2$. Note that, for $N \geq 3$ it is deduced $\beta_p > 0$ for every $N/2 < p \leq \infty$ by applying Lemma 4.3 and the inclusions $L^p \subset L^{N/2}$. The main point here is that these values are attained in all cases. Moreover, we will give a complete information about the functions $a \in \Lambda$ in which β_p is attained. We begin by studying the case $p = \infty$.

Lemma 4.4. β_∞ *is attained in a unique element* $a_\infty \in \Lambda$. *Moreover* $a_\infty(x) \equiv \lambda_1$, *where* λ_1 *is the first strictly positive eigenvalue of the Neumann eigenvalue problem.*

Proof. If $a \in \Lambda \cap L^\infty(\Omega)$ and $u \in H^1(\Omega)$ is a nontrivial solution of (4.1), applying (4.13) yields

$$\int_\Omega |\nabla(u+k)|^2 \le \|a^+\|_\infty \int_\Omega (u+k)^2$$

for every $k \in \mathbf{R}$. As the critical case, choose $k_0 \in \mathbf{R}$ satisfying $u + k_0 \in X_\infty$. Then, since $u + k_0$ is not the trivial function we can asset that

$$\|a^+\|_\infty \ge \frac{\int_\Omega |\nabla(u+k_0)|^2}{\int_\Omega (u+k_0)^2} \ge \inf_{v\in X_\infty \setminus \{0\}} \frac{\int_\Omega |\nabla v|^2}{\int_\Omega v^2} = \lambda_1, \ \forall a \in \Lambda.$$

Hence $\beta_\infty \ge \lambda_1$. Since the constant function λ_1 is an element of Λ, we deduce $\beta_\infty = \lambda_1$. Furthermore, if $a \in \Lambda$ is such that $\|a^+\|_\infty = \lambda_1$, then all the inequalities of the previous proof become equalities. In particular, it follows that

$$\frac{\int_\Omega |\nabla(u+k_0)|^2}{\int_\Omega (u+k_0)^2} = \lambda_1.$$

The variational characterization of λ_1 (this constant is the second eigenvalue of the eigenvalue problem (4.4)) implies that $u(x) + k_0$ is an eigenfunction associated with λ_1. Therefore $-\Delta(u + k_0) = \lambda_1(u + k_0) = a(x)u$. Multiplying by $u + k_0$ we obtain $\int_\Omega (\lambda_1 - a)(u + k_0)^2 \le 0$. Since $\|a^+\|_\infty = \lambda_1$, we deduce $\int_\Omega (\lambda_1 - a)$ $(u + k_0)^2 = 0$. The unique continuation property of the eigenfunctions implies that $u(x) + k_0$ vanishes in a set of measure zero and therefore $a(x) \equiv \lambda_1$. This completes the proof of the lemma.

Next we concentrate on the case $\frac{N}{2} < p < \infty$. We will adapt the techniques of the calculus of β_p in the one-dimensional case to the N-dimensional case. As seen later, the condition $p > N/2$ will be essential: this is equivalent to $2p/(p-1) < 2N/(N-2)$ and hence, by the compact Sobolev embedding, $H^1 \subset L^{2p/(p-1)}$, we will be able to proceed similarly to the ordinary case.

We will need some auxiliary lemmas.

Lemma 4.5. *Assume* $\frac{N}{2} < p < \infty$ *and let* $X_p = \left\{ u \in H^1(\Omega) : \int_\Omega |u|^{\frac{2}{p-1}} u = 0 \right\}$.
If $J_p : X_p \setminus \{0\} \to \mathbf{R}$ *is defined by*

$$J_p(u) = \frac{\int_\Omega |\nabla u|^2}{\left(\int_\Omega |u|^{\frac{2p}{p-1}} \right)^{\frac{p-1}{p}}} \tag{4.14}$$

and $m_p \equiv \inf_{X_p \setminus \{0\}} J_p$, m_p is attained. Moreover, if $u_p \in X_p \setminus \{0\}$ is a minimizer, then u_p satisfies the problem

$$\left. \begin{array}{ll} -\Delta u_p(x) = A_p(u_p)|u_p(x)|^{\frac{2}{p-1}} u_p(x), & x \in \Omega \\ \frac{\partial u_p}{\partial n}(x) = 0, & x \in \partial \Omega \end{array} \right\}, \tag{4.15}$$

where

$$A_p(u_p) = m_p \left(\int_\Omega |u_p|^{\frac{2p}{p-1}} \right)^{\frac{-1}{p}}. \tag{4.16}$$

Proof. It is clear that $X_p \neq \{0\}$ and hence m_p is well defined. Now, let $\{u_n\} \subset X_p \setminus \{0\}$ be a minimizing sequence. Since the sequence $\{k_n u_n\}$, $k_n \neq 0$, is also a minimizing sequence, we can assume without loss of generality that $\int_\Omega |u_n|^{\frac{2p}{p-1}} = 1$. Then $\left\{ \int_\Omega |\nabla u_n|^2 \right\}$ is also bounded. On the other hand, since $p > \frac{N}{2}$, then $2 < \frac{2p}{p-1} < \frac{2N}{N-2}$, which is the critical Sobolev exponent. Moreover the inclusion $H^1 \subset L^{2p/(p-1)}$ is compact. Hence, $\{u_n\}$ is bounded in $H^1(\Omega)$ and we can suppose, up to a subsequence, that $u_n \rightharpoonup u_0$ in $H^1(\Omega)$ and $u_n \to u_0$ in $L^{\frac{2p}{p-1}}$. The strong convergence in $L^{\frac{2p}{p-1}}$ gives us $\int_\Omega |u_0|^{\frac{2p}{p-1}} = 1$ and $u_0 \in X_p \setminus \{0\}$. The weak convergence in $H^1(\Omega)$ implies $J_p(u_0) \leq \liminf J_p(u_n) = m_p$. Then u_0 is a minimizer.

Observe that we can write that $u_p \in X_p \setminus \{0\}$ is a minimizer in an equivalent way: $H(u) \geq H(u_p) = 0$ for every $u \in X_p$, where $H : H^1(\Omega) \to \mathbf{R}$ is defined by

$$H(u) = \frac{1}{2} \int_\Omega |\nabla u|^2 - \frac{1}{2} m_p \left(\int_\Omega |u|^{\frac{2p}{p-1}} \right)^{\frac{p-1}{p}}.$$

In other words, H has a global minimum at $u_p \in X_p = \{ u \in H^1(\Omega) : \varphi(u) = 0 \}$, where

$$\varphi(u) = \int_\Omega |u|^{\frac{2}{p-1}} u.$$

Since $H, \varphi \in C^1(H^1)$ with

$$H'(u_p)(v) = \int_\Omega \nabla u_p \nabla v - m_p \left(\int_\Omega |u_p|^{\frac{2p}{p-1}} \right)^{\frac{-1}{p}} \int_\Omega |u_p|^{\frac{2}{p-1}} u_p\, v \quad \text{for every } v \in H^1,$$

$$\varphi'(u_p)(v) = \frac{p+1}{p-1} \int_\Omega |u_p|^{\frac{2}{p-1}} v \quad \text{for every } v \in H^1.$$

Lagrange multiplier Theorem implies that there is $\lambda \in \mathbf{R}$ such that

$$H'(u_p) + \lambda\varphi'(u_p) \equiv 0.$$

Also, since $u_p \in X_p$ we have $H'(u_p)(1) = 0$. Moreover $H'(u_p)(v) = 0$, $\forall\, v \in H^1(\Omega)$: $\varphi'(u_p)(v) = 0$. Finally, as any $v \in H^1(\Omega)$ may be written in the form $v = \alpha + w$, $\alpha \in \mathbf{R}$, and w satisfying $\varphi'(u_0)(w) = 0$, we conclude $H'(u_p)(v) = 0$, $\forall\, v \in H^1(\Omega)$, i.e., $H'(u_p) \equiv 0$ which is the weak formulation of (4.15).

Lemma 4.6. *If $\frac{N}{2} < p < \infty$, then β_p is attained and $\beta_p = m_p$. Moreover, any function $a \in \Lambda \cap L^p(\Omega)$ in which β_p is attained is of the form $a(x) \equiv |u(x)|^{\frac{2}{p-1}}$, where $u(x)$ is a solution of (4.6).*

Proof. If $a \in \Lambda \cap L^p(\Omega)$ and $u \in H^1(\Omega)$ is a nontrivial solution of (4.1), applying (4.13) and Hölder inequality we obtain

$$\int_\Omega |\nabla(u+k)|^2 \leq \|a^+\|_p\|(u+k)^2\|_{p/(p-1)} = \|a^+\|_p \left(\int_\Omega |u+k|^{\frac{2p}{p-1}}\right)^{\frac{p-1}{p}}$$

for every $k \in \mathbf{R}$.

Also, since u is a nonconstant solution of (4.1), $u + k$ is a nontrivial function. Consequently

$$\|a^+\|_p \geq \frac{\displaystyle\int_\Omega |\nabla(u+k)|^2}{\left(\displaystyle\int_\Omega |u+k|^{\frac{2p}{p-1}}\right)^{\frac{p-1}{p}}}.$$

Now, choose $k_0 \in \mathbf{R}$ satisfying $u + k_0 \in X_p$. Then, $\|a^+\|_p \geq J_p(u+k_0) \geq m_p$, $\forall\, a \in \Lambda \cap L^p(\Omega)$ and consequently $\beta_p \geq m_p$. Conversely, if $u_p \in X_p \setminus \{0\}$ is any minimizer of J_p, then u_p satisfies (4.15). Therefore, $A_p(u_p)|u_p|^{\frac{2}{p-1}} \in \Lambda$. Also,

$$\| A_p(u_p)|u_p|^{\frac{2}{p-1}} \|_p^p = A_p(u_p)^p \int_\Omega |u_p|^{\frac{2p}{p-1}} = m_p^p.$$

Then $\beta_p = m_p$ and β_p is attained.

On the other hand, let $a \in \Lambda \cap L^p(\Omega)$ be such that $\|a^+\|_p = \beta_p$. Then all the inequalities we have used become equalities. In particular, since the above Hölder inequality becomes equality, taking into account (4.13) we have that there exists $M > 0$ such that $a(x) \equiv M|u(x) + k_0|^{\frac{2}{p-1}}$. Hence $a(x) \geq 0$ and consequently $\int_\Omega a > 0$. Therefore, since $\int_\Omega au^2 = \int_\Omega a(u+k_0)^2$ we deduce $k_0 = 0$. Finally, if we define $w(x) = M^{\frac{p-1}{2}} u(x)$ we have that $|w(x)|^{\frac{2}{p-1}} = M|u(x)|^{\frac{2}{p-1}} = a(x)$. Moreover, since $u(x)$ is a solution of (4.1) and $w(x)$ is a multiple of $u(x)$, then also $w(x)$ is a solution of (4.1) and consequently a solution of (4.6), and the lemma follows.

4.2.4 Qualitative Properties of β_p

In this subsection we will study some qualitative aspects of the function $p \mapsto \beta_p$. Specifically, we will prove some results of continuity, monotonicity, and behavior of β_p when p is near $\frac{N}{2}$ and $+\infty$.

Proofs of 3 and 4 of Theorem 4.1. We first prove the continuity of β_p in $(\frac{N}{2}, \infty)$. To this aim, consider a sequence $\{p_n\} \to p \in (\frac{N}{2}, \infty)$. Take a function $a_p \in \Lambda \cap L^p(\Omega)$ such that $\|a_p^+\|_p = \beta_p$. By (2) of Theorem 4.1, and using standard regularity arguments, we have $a_p \in L^\infty(\Omega)$. Hence $\|a_p\|_{p_n} \to \|a_p\|_p$ and it follows that

$$\limsup \beta_{p_n} \leq \limsup \|a_p\|_{p_n} = \|a_p\|_p = \beta_p.$$

In order to obtain the inverse inequality, and using that $\beta_p = m_p$, consider a nonzero sequence $\{u_{p_n}\} \subset X_{p_n} = \left\{ u \in H^1(\Omega) : \int_\Omega |u|^{\frac{2}{p_n-1}} u = 0 \right\}$ and $J_{p_n}(u_{p_n}) = \beta_{p_n}$. We can suppose without loss of generality that $\|u_{p_n}\|_{\frac{2p_n}{p_n-1}} = 1$. Consequently $\|u_{p_n}\|_2$ is bounded. On the other hand, $\int_\Omega |\nabla u_{p_n}|^2 = \beta_{p_n}$, which is a bounded sequence, from the above inequality. Therefore, $\{u_{p_n}\}$ is bounded in $H^1(\Omega)$ and, up to a subsequence, $\{u_{p_n}\} \rightharpoonup u_0$ in $H^1(\Omega)$ for some $u_0 \in H^1(\Omega)$. Take $q < \frac{2N}{N-2}$, such that $2p/(p-1) < q < 2N/(N-2)$. The strong convergence of u_{p_n} to u_0 in L^q shows that

$$\liminf \beta_{p_n} = \liminf \frac{\int_\Omega |\nabla u_{p_n}|^2}{\left(\int_\Omega |u_{p_n}|^{\frac{2p_n}{p_n-1}} \right)^{\frac{p_n-1}{p_n}}} \geq \frac{\int_\Omega |\nabla u_0|^2}{\left(\int_\Omega |u_0|^{\frac{2p}{p-1}} \right)^{\frac{p-1}{p}}} \geq \beta_p$$

and the continuity of β_p is proved.

We now prove that the mapping $[\frac{N}{2}, \infty) \to \mathbf{R}$, $p \mapsto |\Omega|^{-1/p}\beta_p$ is strictly increasing in $[\frac{N}{2}, \infty)$. To do this, take $\frac{N}{2} \leq q < p < \infty$. Taking into account that $|\Omega|^{-1/q}\|f\|_q \leq |\Omega|^{-1/p}\|f\|_p$ for every $f \in L^p(\Omega)$ (strict inequality if $|f|$ is not constant) we have

$$|\Omega|^{-1/q}\beta_q \leq |\Omega|^{-1/q}\|a_p\|_q \leq |\Omega|^{-1/p}\|a_p\|_p = |\Omega|^{-1/p}\beta_p.$$

Since $|a_p|$ is not constant, we have that the above inequality is strict.

On the other hand, similar arguments of the continuity of β_p in $(\frac{N}{2}, \infty)$ gives us $\lim_{p\to\infty} \beta_p = \beta_\infty$.

To study the behavior of β_p, for p near $\frac{N}{2}$, let us observe that since $|\Omega|^{-1/p}\beta_p$ is strictly increasing in $[\frac{N}{2}, \infty)$, then

$$\exists \lim_{p \to \frac{N}{2}^+} |\Omega|^{-1/p}\beta_p \geq |\Omega|^{\frac{-1}{N/2}} \beta_{\frac{N}{2}}.$$

Hence $\exists \lim_{p \to \frac{N}{2}^+} \beta_p \geq \beta_{\frac{N}{2}} > 0$, if $N \geq 3$.

Finally, let us consider the case $N = 2$. If we fixed a function $a \in \Lambda \cap L^\infty(\Omega)$ we have

$$\limsup_{p \to 1^+} \beta_p \leq \limsup_{p \to 1^+} \|a^+\|_p = \|a^+\|_1.$$

But, when we prove $\beta_1 = 0$, for $N = 2$, we have used nonnegative minimizing functions $a \in \Lambda \cap L^\infty(\Omega)$. Then we can conclude

$$\lim_{p \to 1^+} \beta_p = 0.$$

$\square$

4.3   Nonlinear Problems

In this section we give some results on the existence and uniqueness of solutions of nonlinear boundary value problems of the type

$$\left.\begin{array}{ll} -\Delta u(x) = f(x, u(x)), & x \in \Omega \\ \frac{\partial u}{\partial n}(x) = 0, & x \in \partial\Omega \end{array}\right\}, \tag{4.17}$$

where $\Omega \subset \mathbf{R}^N$ ($N \geq 2$) is a bounded and regular domain and the function $f : \overline{\Omega} \times \mathbf{R} \to \mathbf{R}$, $(x, u) \mapsto f(x, u)$, satisfies certain conditions.

As we will see, Theorem 4.2 is a generalization of Theorem 2 in [8] in the sense that the main hypothesis of $f(x, u)$ in [8] is given in terms of an L^∞-restriction, while we give here a more general L^p-restriction for $\frac{N}{2} < p \leq \infty$. In the proof, the basic idea is to combine the results obtained in the previous section with the Schauder's fixed point theorem. In fact, once we have the results on the linear problem, the procedure is standard and may be seen, for example, in [1, 7].

The following lemma follows easily from Theorem 4.1 and will be useful in order to prove Theorem 4.2.

Lemma 4.7. *Let* $a \in L^p(\Omega) \setminus \{0\}$ *(for some* $p > N/2$*),* $0 \leq \int_\Omega a(x)$*, satisfying* $\|a^+\|_p < \beta_p$ *(or* $\|a^+\|_p = \beta_p$ *and* $a(x)$ *is not a minimizer of the* L_p*-norm in* Λ*). Then for each* $f \in L^p(\Omega)$ *the boundary value problem*

$$\left.\begin{array}{ll} -\Delta u(x) = a(x)u(x) + f(x), & x \in \Omega \\ \frac{\partial u}{\partial n}(x) = 0, & x \in \partial\Omega \end{array}\right\}$$

has a unique solution.

Theorem 4.2. *Let* $\Omega \subset \mathbf{R}^N$ ($N \geq 2$) *be a bounded and regular domain and* $f : \overline{\Omega} \times \mathbf{R} \to \mathbf{R}$, $(x, u) \mapsto f(x, u)$, *satisfying:*

1. f, f_u *are Caratheodory functions and* $f(\cdot, 0) \in L^p(\Omega)$ *for some* $\frac{N}{2} < p \leq \infty$.
2. *There exist functions* $\alpha, \beta \in L^p(\Omega)$, *satisfying*

$$\alpha(x) \leq f_u(x, u) \leq \beta(x) \text{ in } \overline{\Omega} \times \mathbf{R} \tag{4.18}$$

with $\|\beta^+\|_p < \beta_p$ (or $\|\beta^+\|_p = \beta_p$ and $\beta(x)$ is not a minimizer of the L_p-norm in Λ), where β_p is given by Theorem 4.1.
Moreover, we assume one of the following conditions

(a)

$$\int_\Omega \alpha \geq 0, \ \alpha \not\equiv 0 \tag{4.19}$$

(b)

$$\alpha \equiv 0, \ \exists s_0 \in \mathbf{R} \ s.t. \ \int_\Omega f(x, s_0) \, dx = 0, \ and \ f_u(x, u(x)) \not\equiv 0, \ \forall u \in C(\overline{\Omega}) \tag{4.20}$$

Then problem (4.17) has a unique solution.

Proof. We first prove uniqueness. Let u_1 and u_2 be two solutions of (4.17). Then, the function $u = u_1 - u_2$ is a solution of the problem

$$-\Delta u(x) = a(x)u(x), \ x \in \Omega, \quad \frac{\partial u}{\partial n} = 0, \ x \in \partial\Omega, \tag{4.21}$$

where $a(x) = \int_0^1 f_u(x, u_2(x) + \theta u(x)) \, d\theta$. Hence $\alpha(x) \leq a(x) \leq \beta(x)$ and we deduce $a(x) \in \Lambda$ and $\|a^+\|_p \leq \|\beta\|_p$. Applying Theorem 4.1, we obtain $u \equiv 0$.

Next we prove existence. First, we write (4.17) in the equivalent form

$$\begin{cases} -\Delta u(x) = b(x, u(x))u(x) + f(x, 0), & \text{in } \Omega, \\ \frac{\partial u}{\partial n} = 0, & \text{on } \partial\Omega \end{cases}, \tag{4.22}$$

where the function $b : \Omega \times \mathbf{R} \to \mathbf{R}$ is defined by $b(x, z) = \int_0^1 f_u(x, \theta z) \, d\theta$. Hence $\alpha(x) \leq b(x, z) \leq \beta(x), \ \forall(x, z) \in \Omega \times \mathbf{R}$ and our hypothesis permit us to apply Lemma 4.7 in order to have a well-defined operator $T : X \to X$, by $Ty = u_y$, being u_y the unique solution of the linear problem

$$\begin{cases} -\Delta u(x) = b(x, y(x))u(x) + f(x, 0), & \text{in } \Omega, \\ \frac{\partial u}{\partial n} = 0, & \text{on } \partial\Omega. \end{cases}, \tag{4.23}$$

where $X = C(\overline{\Omega})$ with the uniform norm.

We will show that T is completely continuous and that $T(X)$ is bounded. The Schauder's fixed point theorem provides a fixed point for T which is a solution of (4.17).

The fact that T is completely continuous is a consequence of the compact embedding of the Sobolev space $W^{2,p}(\Omega) \subset C(\overline{\Omega})$. It remains to prove that $T(X)$ is bounded. Suppose, contrary to our claim, that $T(X)$ is not bounded. In this case, there would exist a sequence $\{y_n\} \subset X$ such that $\|u_{y_n}\|_X \to \infty$. Passing to a subsequence if necessary, we may assume that the sequence of functions $\{b(\cdot, y_n(\cdot))\}$ is weakly convergent in $L^p(\Omega)$ to a function a_0 satisfying $\alpha(x) \le a_0(x) \le \beta(x)$, a.e. in Ω. If $z_n \equiv \dfrac{u_{y_n}}{\|u_{y_n}\|_X}$, passing to a subsequence if necessary, we may assume that $z_n \to z_0$ strongly in X (we have used again the compact embedding $W^{2,p}(\Omega) \subset C(\overline{\Omega})$), where z_0 is a nonzero function satisfying

$$\left.\begin{array}{ll} -\Delta z_0(x) = a_0(x)z_0(x), & \text{in } \Omega, \\ \frac{\partial z_0}{\partial n} = 0, & \text{on } \partial\Omega \end{array}\right\}. \tag{4.24}$$

Now, if we are assuming (4.19), then $a_0 \not\equiv 0$, $a_0 \in \Lambda$ and we obtain a contradiction with Theorem 4.1. If we are assuming (4.20), there is no loss of generality if we suppose $s_0 = 0$. (Otherwise, we can do the change of variables $u(x) = v(x) + s_0$ and obtain a similar problem with the same original hypothesis). Then for every $n \in \mathbf{N}$, $\displaystyle\int_\Omega b(x, y_n(x))u_{y_n}(x) \, dx = -\int_\Omega f(x, 0) \, dx = 0$. Therefore, for each $n \in \mathbf{N}$, the function u_{y_n} has a zero in $\overline{\Omega}$ and hence so does z_0. Thus, $a_0 \not\equiv 0$, $a_0 \in \Lambda$ and we obtain again a contradiction with Theorem 4.1 $\square$

4.4 Radial Higher Eigenvalues

In the previous sections of this chapter we have done a careful study of Lyapunov inequalities for partial differential equations. One can ask if it is possible to do an analysis for higher eigenvalues, proceeding similarly as in Chap. 3 for ordinary differential equations. In Chap. 3 a detailed analysis about the number and distribution of zeros of nontrivial solutions and their first derivatives is essential to obtain the main results. Unfortunately, it seems very difficult to do an analogous study of the nodal sets of N-dimensional solutions. However, if we restrict our attention to a class of solutions of P.D.Es, the radial functions, it is possible to obtain interesting results. Obviously the nodal sets of radial solutions are spheres of different radii. This fact, together with appropriate minimizing sequences, will allow us to obtain L_p Lyapunov inequalities. Again the relation between the quantities p and $N/2$ plays a crucial role.

Through this section, we will consider $\Omega = B_1$, the open ball in $\mathbf{R}^N$ of center zero and radius one. It is very well known [3] that the operator $-\Delta$ exhibits an infinite increasing sequence of radial Neumann eigenvalues $0 = \mu_0 < \mu_1 < \ldots < \mu_k < \ldots$ with $\mu_k \to +\infty$, all of them simple and with associated eigenfunctions $\varphi_k \in C^1[0, 1]$ solving

$$-(r^{N-1}\varphi')' = \mu_k r^{N-1}\varphi, \ 0 < r < 1$$
$$\varphi'(1) = 0. \tag{4.25}$$

Moreover, each eigenfunction φ_k has exactly k simple zeros $r_k < r_{k-1} < \ldots < r_1$ in the interval $(0, 1)$.

For each integer $k \geq 0$ and number p, $1 \leq p \leq \infty$, we can define the set

$$\Gamma_k = \{a \in L^{N/2}(B_1) : \ a \text{ is a radial function, } \mu_k \prec a \text{ and}$$

(4.1) has radial and nontrivial solutions $\}$ if $N \geq 3$ and

$$\Gamma_k = \{a : B_1 \to \mathbf{R} \text{ s. t. } \exists q \in (1, \infty] \text{ with } a \in L^q(B_1) : \ a \text{ is a radial function,}$$

$$\mu_k \prec a \text{ and (4.1) has radial and nontrivial solutions}\}$$

if $N = 2$.

We also define the quantity

$$\gamma_{p,k} = \inf_{a \in \Gamma_k \cap L^p(B_1)} \|a - \mu_k\|_{L^p(B_1)}. \tag{4.26}$$

The main result of this section is the following.

Theorem 4.3. *Let $k \geq 0$, $N \geq 2$, $1 \leq p \leq \infty$. The following statements hold:*

1. *If $N = 2$, then $\gamma_{p,k} > 0 \Leftrightarrow 1 < p \leq \infty$.*
   If $N \geq 3$, then $\gamma_{p,k} > 0 \Leftrightarrow \frac{N}{2} \leq p \leq \infty$.
2. *If $\frac{N}{2} < p \leq \infty$, then $\gamma_{p,k}$ is attained.*

A key ingredient to prove this theorem is the following proposition on the number and distribution of zeros of nontrivial radial solutions of (4.1) when $a \in \Gamma_k$.

Proposition 4.1. *Let $\Omega = B_1$, $k \geq 0$, $a \in \Gamma_k$ and u any nontrivial radial solution of (4.1). Then u has, at least, $k + 1$ zeros in $(0, 1)$. Moreover, if $k \geq 1$ and we denote by $x_k < x_{k-1} < \ldots < x_1$ the last k zeros of u, we have that*

$$r_i \leq x_i, \ \forall \ 1 \leq i \leq k,$$

where r_i denotes the zeros of the eigenfunction φ_k of (4.25).

First of all, observe that from the uniqueness of the corresponding Cauchy problems, all the zeros in $(0, 1)$ of nontrivial radial solutions are isolated. Therefore it has sense to consider the "last k zeros of u."

For the proof of this proposition we will need the following lemma. Some of the results of this lemma can be proved in a different way, by using the version of the Sturm Comparison Lemma proved in [3], Lemma 4.1, for the p-laplacian operator (see also [6]).

Lemma 4.8. *Let* $k \geq 1$. *Under the hypothesis of Proposition 4.1 we have that*

i) *u admits a zero in the interval* $(0, r_k]$. *If* r_k *is the only zero of u in this interval, then* $a(r) \equiv \mu_k$ *in* $(0, r_k]$.

ii) *u admits a zero in the interval* $[r_{i+1}, r_i)$, *for* $1 \leq i \leq k-1$. *If* r_{i+1} *is the only zero of u in this interval, then* $u(r_i) = 0$ *and* $a(r) \equiv \mu_k$ *in* $[r_{i+1}, r_i]$.

ii) *u admits a zero in the interval* $[r_1, 1)$. *If* r_1 *is the only zero of u in this interval, then* $a(r) \equiv \mu_k$ *in* $[r_1, 1]$.

Proof. To prove i), multiplying (4.1) by φ_k and integrating by parts in B_{r_k} (the ball centered in the origin of radius r_k), we obtain

$$\int_{B_{r_k}} \nabla u \nabla \varphi_k = \int_{B_{r_k}} a u \varphi_k.$$

On the other hand, multiplying (4.25) by u and integrating by parts in B_{r_k}, we have

$$\int_{B_{r_k}} \nabla \varphi_k \nabla u = \mu_k \int_{B_{r_k}} \varphi_k u + \int_{\partial B_{r_k}} u \frac{\partial \varphi_k}{\partial n}.$$

Subtracting these equalities yields

$$\int_{B_{r_k}} (a - \mu_k) u \varphi_k = \omega_N r_k^{N-1} u(r_k) \varphi_k'(r_k). \tag{4.27}$$

Assume, by contradiction, that u does not admit any zero in $(0, r_k]$. We can suppose, without loss of generality, that $u > 0$ in this interval. We can also assume that $\varphi_k > 0$ in $(0, r_k)$. Since r_k is a simple zero of φ_k, we have $\varphi_k'(r_k) < 0$ and since $a \geq \mu_k$ in $(0, r_k)$ we obtain a contradiction.

Finally, if r_k is the only zero of u in $(0, r_k]$, Eq. (4.27) yields $\int_{B_{r_k}} (a - \mu_k) u \varphi_k = 0$, which gives $a(r) \equiv \mu_k$ in $(0, r_k]$.

To deduce ii), we proceed similarly to the proof of part i), restituting B_{r_k} by $A(r_{i+1}, r_i)$ (the annulus centered in the origin of radii r_{i+1} and r_i) and obtaining

$$\int_{A(r_{i+1}, r_i)} (a - \mu_k) u \varphi_k = \omega_N r_i^{N-1} u(r_i) \varphi_k'(r_i) - \omega_N r_{i+1}^{N-1} u(r_{i+1}) \varphi_k'(r_{i+1})$$

and ii) follows easily by arguments on the sign of these quantities, as in the proof of part i).

To obtain iii), a similar analysis to that in the previous cases shows that

$$\int_{A(r_i, 1)} (a - \mu_k) u \varphi_k = -\omega_N r_1^{N-1} u(r_1) \varphi_k'(r_1)$$

and the lemma follows easily as previously.

Proof of Proposition 4.1. Let $k = 0$. If we suppose that u has no zeros in $(0, 1]$ and we integrate the equation $-\Delta u = a u$ in B_1, we obtain $\int_{B_1} a u = 0$, a contradiction. Hence, for the rest of the proof we will consider $k \geq 1$.

Let $1 \leq i \leq k$. By the previous lemma u admits a zero in each of i disjoint intervals $[r_i, r_{i-1}), \ldots, [r_2, r_1), [r_1, 1)$. Therefore u has, at least, i zeros in the interval $[r_i, 1)$ which implies that $r_i \leq x_i$.

Finally, let us prove that u has, at least, $k + 1$ zeros. From the previous part, taking $i = k$, u has at least k zeros in the interval $[r_k, 1]$, one in each of the k disjoint intervals $[r_k, r_{k-1}), \ldots, [r_2, r_1), [r_1, 1)$. Suppose, by contradiction, that these are the only zeros of u. Then u does not admit any zero in $(0, r_k)$ and applying part (i) of Lemma 4.8 we obtain $u(r_k) = 0$ and $a \equiv \mu_k$ in $(0, r_k]$. Applying now part (ii) of this lemma, we deduce $u(r_{k-1}) = 0$ and $a \equiv \mu_k$ in $[r_k, r_{k-1}]$. Repeating this argument and using part (iii) of the previous lemma we conclude $u(r_i) = 0$, for all $1 \leq i \leq k$ and $a \equiv \mu_k$ in $(0,1]$, which contradicts $a \in \Gamma_k$.   □

For the proof of Theorem 4.3, we will distinguish three cases: the subcritical case ($1 \leq p < \frac{N}{2}$ if $N \geq 3$, and $p = 1$ if $N = 2$), the critical case ($p = \frac{N}{2}$ if $N \geq 3$), and the supercritical case ($p > \frac{N}{2}$ if $N \geq 2$).

4.4.1 The Subcritical Case

In this subsection, we study the subcritical case, i.e., $1 \leq p < \frac{N}{2}$, if $N \geq 3$, and $p = 1$ if $N = 2$. In all these cases we will prove that $\gamma_{p,k} = 0$.

The next lemma is related to the continuous domain dependence of the eigenvalues of the Dirichlet Laplacian. In fact, the result is valid under much more general hypothesis (see [5]). Here we show a very simple proof for this special case.

Lemma 4.9. *Let $N \geq 2$ and $R > 0$. Then*

$$\lim_{\varepsilon \to 0} \lambda_1 (A(\varepsilon, R)) = \lambda_1 (B_R),$$

where $\lambda_1 (A(\varepsilon, R))$ and $\lambda_1 (B_R)$ denotes, respectively, the first eigenvalue of the Laplacian operator with Dirichlet boundary conditions of the annulus $A(\varepsilon, R)$ and the ball B_R.

Proof. For $N \geq 3$ and $\varepsilon \in (0, R/2)$ define the following radial function $u_\varepsilon \in H_0^1 (A(\varepsilon, R))$:

$$u_\varepsilon(x) = \begin{cases} \phi_1(x), & \text{if } 2\varepsilon \leq |x| < R \\ \dfrac{|x| - \varepsilon}{\varepsilon} \phi_1(2\varepsilon), & \text{if } \varepsilon < |x| < 2\varepsilon, \end{cases} \tag{4.28}$$

where ϕ_1 denotes the first eigenfunction with Dirichlet boundary conditions of the ball B_R. It is easy to check that

$$\lim_{\varepsilon \to 0} \int_{A(\varepsilon, 2\varepsilon)} |\nabla u_\varepsilon|^2 = \lim_{\varepsilon \to 0} \int_\varepsilon^{2\varepsilon} \omega_N r^{N-1} \frac{\phi_1(2\varepsilon)^2}{\varepsilon^2} dr = 0.$$

In the same way it is obtained $\lim_{\varepsilon \to 0} \int_{A(\varepsilon, 2\varepsilon)} u_\varepsilon^2 = 0$. In addition, from the variational characterization of the first eigenvalue it follows that $\lambda_1 (A(\varepsilon, R)) \leq \dfrac{\displaystyle\int_{A(\varepsilon, R)} |\nabla u_\varepsilon|^2}{\displaystyle\int_{A(\varepsilon, R)} u_\varepsilon^2}$

Therefore

$$\limsup_{\varepsilon \to 0} \lambda_1 (A(\varepsilon, R)) \leq \limsup_{\varepsilon \to 0} \frac{\int_{A(\varepsilon, R)} |\nabla u_\varepsilon|^2}{\int_{A(\varepsilon, R)} u_\varepsilon^2} = \frac{\int_{B_R} |\nabla \phi_1|^2}{\int_{B_R} \phi_1^2} = \lambda_1(B_R).$$

On the other hand, using that the first Dirichlet eigenvalue $\lambda_1(\Omega)$ is strictly decreasing with respect to the domain Ω, it follows that $\lambda_1 (A(\varepsilon, R)) > \lambda_1 (B_R)$. Thus

$$\liminf_{\varepsilon \to 0} \lambda_1 (A(\varepsilon, R)) \geq \lambda_1 (B_R)$$

and the lemma follows for $N \geq 3$.

The same proof works for $N = 2$ if we consider, for every $\varepsilon \in (0, \min\{1, R^2\})$, the radial function $u_\varepsilon \in H_0^1 (A(\varepsilon, R))$:

$$u_\varepsilon(x) = \begin{cases} \phi_1(x), & \text{if } \sqrt{\varepsilon} \leq |x| < R \\ \dfrac{\log |x| - \log \varepsilon}{\log \sqrt{\varepsilon} - \log \varepsilon} \phi_1(\sqrt{\varepsilon}), & \text{if } \varepsilon < |x| < \sqrt{\varepsilon}. \end{cases} \tag{4.29}$$

Lemma 4.10. *Let $k \geq 0$, $N \geq 3$ and $1 \leq p < N/2$. Then $\gamma_{p,k} = 0$.*

Proof. If $k = 0$, this lemma follows from Lemma 4.1. In this lemma a family of nonnegative and radial functions was used. Hence, for the rest of the proof we will consider $k \geq 1$.

To prove this lemma we will construct an explicit family $a_\varepsilon \in \Gamma_k$ such that $\lim_{\varepsilon \to 0} \|a_\varepsilon - \mu_k\|_{L^p(B_1)} = 0$. To this end, for every $\varepsilon \in (0, r_k)$, define $u_\varepsilon : B_1 \to \mathbf{R}$ as the radial function

$$u_\varepsilon = \begin{cases} \varphi_k, & \text{if } r_k \leq |x| < 1 \\ \phi_1 (A(\varepsilon, r_k)), & \text{if } \varepsilon \leq |x| < r_k \\ \phi_1 (B_\varepsilon), & \text{if } |x| < \varepsilon, \end{cases} \tag{4.30}$$

where $\phi_1(A(\varepsilon, r_k))$ and $\phi_1(B_\varepsilon)$ denotes, respectively, the first eigenfunction with Dirichlet boundary conditions of the annulus $A(\varepsilon, r_k)$ and the ball B_ε. Moreover these eigenfunctions are chosen such that $u_\varepsilon \in C^1(\overline{B_1})$.

Then, it is easy to check that u_ε is a solution of (4.1), being $a_\varepsilon \in L^\infty(B_1)$ the radial function

$$
a_\varepsilon = \begin{cases} \mu_k, & \text{if } r_k < |x| < 1 \\ \lambda_1(A(\varepsilon, r_k)), & \text{if } \varepsilon < |x| < r_k \\ \lambda_1(B_\varepsilon), & \text{if } |x| < \varepsilon, \end{cases} \tag{4.31}
$$

where $\lambda_1(A(\varepsilon, r_k))$ and $\lambda_1(B_\varepsilon)$ denotes, respectively, the first eigenvalues with Dirichlet boundary conditions of the annulus $A(\varepsilon, r_k)$ and the ball B_ε. Since the first Dirichlet eigenvalue $\lambda_1(\Omega)$ is strictly decreasing with respect to the domain Ω, it follows that

$$
\lambda_1(A(\varepsilon, r_k)), \lambda_1(B_\varepsilon) > \lambda_1(B_{r_k}) = \mu_k
$$

which gives $a_\varepsilon \in \Gamma_k$. (The equality $\lambda_1(B_{r_k}) = \mu_k$ follows from the fact that φ_k is a positive solution of $-\Delta\varphi = \mu_k\varphi$ in B_{r_k} which is identically zero on ∂B_{r_k}). Let us estimate the L_p-norm of $a_\varepsilon - \mu_k$:

$$
\|a_\varepsilon - \mu_k\|_{L^p(B_1)} = \left(\int_{B_\varepsilon} (\lambda_1(B_\varepsilon) - \mu_k)^p + \int_{A(\varepsilon, r_k)} (\lambda_1(A(\varepsilon, r_k)) - \mu_k)^p \right)^{\frac{1}{p}}
$$

$$
= \left((\lambda_1(B_\varepsilon) - \mu_k)^p \frac{\omega_N \varepsilon^N}{N} + (\lambda_1(A(\varepsilon, r_k)) - \mu_k)^p \frac{\omega_N(r_k^N - \varepsilon^N)}{N} \right)^{\frac{1}{p}}.
$$
$$\tag{4.32}$$

Taking into account that $\lambda_1(B_\varepsilon) = \lambda_1(B_1)/\varepsilon^2$, $\lambda_1(B_{r_k}) = \mu_k$, using $N > 2p$, and applying Lemma 4.9, we conclude

$$
\lim_{\varepsilon \to 0} \|a - \mu_k\|_{L^p(B_1)} \le \lim_{\varepsilon \to 0} \left(\frac{\lambda_1(B_1)^p}{\varepsilon^{2p}} \frac{\omega_N \varepsilon^N}{N} + (\lambda_1(A(\varepsilon, r_k)) - \mu_k)^p \frac{\omega_N(r_k^N - \varepsilon^N)}{N} \right)^{\frac{1}{p}}
$$

$= 0$ and the proof is complete.

Lemma 4.11. *Let $k \ge 0$, $N = 2$ and $p = 1$. Then $\gamma_{1,k} = 0$.*

Proof. If $k = 0$, this lemma follows from Lemma 4.2. In this lemma a family of nonnegative and radial functions were used. Hence, for the rest of the proof we will consider $k \ge 1$.

Similar to the proof of the previous lemma, we will construct some explicit sequences in Γ_k. In this case, this construction will be slightly more complicated. First, for every $\alpha \in (0, 1)$, define $v_\alpha, A_\alpha : B_1 \to \mathbf{R}$ as the radial functions:

$$v_\alpha(r) = \begin{cases} \alpha(1 - r^2)(3 - r^2) - \log r, & \text{if } \alpha \leq r < 1 \\ \\ \alpha(1 - r^2)(3 - r^2) - \log \alpha + \dfrac{\alpha^2 - r^2}{2\alpha^2}, & \text{if } r < \alpha \end{cases} \tag{4.33}$$

$$A_\alpha(r) = \begin{cases} \dfrac{16\alpha(1 - r^2)}{\alpha(1 - r^2)(3 - r^2) - \log r}, & \text{if } \alpha < r < 1 \\ \\ \dfrac{16\alpha(1 - r^2) + \frac{2}{\alpha^2}}{\alpha(1 - r^2)(3 - r^2) - \log \alpha + \dfrac{\alpha^2 - r^2}{2\alpha^2}}, & \text{if } r < \alpha, \end{cases} \tag{4.34}$$

where $r = |x|$. It is easily seen that $v_\alpha \in C^1(\overline{B_1})$, $A_\alpha \in L^\infty(B_1)$, and

$$\left. \begin{array}{ll} \Delta v_\alpha(x) + A_\alpha(x)v_\alpha(x) = 0, & x \in B_1 \\ v_\alpha(x) = 0, & x \in \partial B_1 \end{array} \right\} . \tag{4.35}$$

Now, for every $\alpha \in (0, 1)$ and $\varepsilon \in (0, r_k)$, define $u_{\alpha,\varepsilon} : B_1 \to \mathbf{R}$ as the radial function:

$$u_{\alpha,\varepsilon}(x) = \begin{cases} \varphi_k(x), & \text{if } r_k \leq |x| < 1 \\ \\ \phi_1\left(A(\varepsilon, r_k)\right)(x), & \text{if } \varepsilon \leq |x| < r_k \\ \\ v_\alpha\left(\dfrac{x}{\varepsilon}\right), & \text{if } |x| < \varepsilon, \end{cases} \tag{4.36}$$

where the eigenfunctions φ_k and $\phi_1\left(A(\varepsilon, r_k)\right)$ are chosen such that $u_{\alpha,\varepsilon} \in C^1(\overline{B_1})$.

An easy computation shows that $u_{\alpha,\varepsilon}$ is a solution of (4.1), being $a_{\alpha,\varepsilon} \in L^\infty(B_1)$ the radial function

$$a_{\alpha,\varepsilon}(x) = \begin{cases} \mu_k, & \text{if } r_k < |x| < 1 \\ \\ \lambda_1\left(A(\varepsilon, r_k)\right), & \text{if } \varepsilon < |x| < r_k \\ \\ \dfrac{1}{\varepsilon^2} A_\alpha\left(\dfrac{x}{\varepsilon}\right), & \text{if } |x| < \varepsilon. \end{cases} \tag{4.37}$$

Again, using that the first Dirichlet eigenvalue $\lambda_1(\Omega)$ is strictly decreasing with respect to the domain Ω, it follows that

$$\lambda_1\left(A(\varepsilon, r_k)\right) > \lambda_1(B_{r_k}) = \mu_k.$$

Moreover, $\inf\limits_{|x|<\varepsilon} a_{\alpha,\varepsilon}(x) = \left(\inf\limits_{x\in B_1} A_\alpha(x)\right)/\varepsilon^2 := m_\alpha/\varepsilon^2$. We see at once that $m_\alpha > 0$ for every $\alpha \in (0,1)$. Hence, if we fix α and choose $\varepsilon \in (0,1)$ such that $m_\alpha/\varepsilon^2 \geq \mu_k$, it is deduced that $a_{\alpha,\varepsilon} \in \Gamma_k$.

Let us estimate the L_1-norm of $a_{\alpha,\varepsilon} - \mu_k$:

$$\|a_{\alpha,\varepsilon} - \mu_k\|_{L^1(B_1)} = \int_{B_\varepsilon}\left(\frac{1}{\varepsilon^2}A_\alpha\left(\frac{x}{\varepsilon}\right) - \mu_k\right)dx + \int_{A(\varepsilon,r_k)}(\lambda_1(A(\varepsilon,r_k)) - \mu_k)\,dx. \tag{4.38}$$

Doing the change of variables $x = \varepsilon y$ in the first integral and applying Lemma 4.9 in the second one, it is obtained, for fixed $\alpha \in (0,1)$:

$$\lim_{\varepsilon\to 0}\|a_{\alpha,\varepsilon} - \mu_k\|_{L^1(B_1)} = \int_{B_1}A_\alpha(y)dy.$$

Thus, from the definition of $\gamma_{1,k}$ we have

$$\gamma_{1,k} \leq \int_{B_1}A_\alpha(y)dy, \quad \forall\alpha \in (0,1). \tag{4.39}$$

Now we will take limit when α tends to 0 in this last expression. For this purpose we first deduce easily from the definition of A_α that $A_\alpha(r) \leq 16\alpha(1 - r^2)/(-\log r) \leq 32\alpha$ if $r \in (\alpha, 1)$ and $A_\alpha(r) \leq \left(16\alpha + 2/\alpha^2\right)/(-\log\alpha)$ if $r \in (0,\alpha)$. It follows that

$$\int_{B_1}A_\alpha(y)dy = 2\pi\int_0^1 rA_\alpha(r)dr \leq 2\pi\int_0^\alpha r\frac{16\alpha + 2/\alpha^2}{-\log\alpha}dr + 2\pi\int_\alpha^1 r\,32\alpha dr$$

$$= \pi\frac{16\alpha^3 + 2}{-\log\alpha} + 32\pi\alpha(1 - \alpha^2)$$

which gives $\lim_{\alpha\to 0}\int_{B_1}A_\alpha(y)dy = 0$ and the lemma follows from (4.39).

4.4.2 The Critical Case

In this subsection, we study the critical case, i.e., $p = \frac{N}{2}$, if $N \geq 3$. We will prove that $\gamma_{p,k} > 0$.

Lemma 4.12. *Let $k \geq 0$, $N \geq 3$, and $p = N/2$. Then $\gamma_{p,k} > 0$.*

Proof. To obtain a contradiction, suppose that $\gamma_{p,k} = 0$. Then we could find a sequence $\{a_n\} \subset \Gamma_k$ such that $a_n \to \mu_k$ in $L^{N/2}(B_1)$ and a sequence $\{u_n\} \subset H^1(B_1)$ such that u_n is a radial solution of (4.1), for $a = a_n$, with the normalization

$\|u_n\|^2_{H^1(B_1)} = \int_{B_1} (|\nabla u_n|^2 + u_n^2) = 1$. We can suppose, up to a subsequence, that $u_n \rightharpoonup u_0$ in $H^1(B_1)$ and taking limits in Eq. (4.1), for $a = a_n$ and $u = u_n$, we obtain that u_0 is a solution of this equation for $a = \mu_k$.

We claim that $u_n \to u_0$ in $H^1(B_1)$ and consequently, $u_0 = \varphi_k$, for some nontrivial eigenfunction φ_k. For this purpose, we set

$$\lim \int_{B_1} |\nabla u_n|^2 = \lim \int_{B_1} a_n u_n^2 = \lim \int_{B_1} (a_n - \mu_k)u_n^2 + \lim \int_{B_1} \mu_k u_n^2$$

$$= 0 + \mu_k \int_{B_1} u_0^2 = \int_{B_1} |\nabla u_0|^2,$$

where we have used $a_n \to \mu_k$ in $L^{N/2}(B_1)$ and u_n^2 is bounded in $L^{N/(N-2)}(B_1)$ (since u_n is bounded in $H^1(B_1) \subset L^{2N/(N-2)}(B_1)$). Thus, from standard arguments, we deduce that $u_n \to u_0 = \varphi_k$ in $H^1(B_1)$.

In the following, we will fix $\varepsilon \in (0, r_k)$. Since $a_n \to \mu_k$ in $L^{N/2}(A(\varepsilon, 1))$ and $u_n \to u_0 = \varphi_k$ in $H^1_{\text{rad}}(A(\varepsilon, 1)) \subset C(A(\varepsilon, 1))$, we can assert that $a_n u_n \to \mu_k \varphi_k$ in $L^{N/2}(A(\varepsilon, 1)) \subset L^1(A(\varepsilon, 1))$. Thus $-\Delta u_n \to \mu_k \varphi_k$ in $L^1(A(\varepsilon, 1))$, which yields $u_n \to \varphi_k$ in $C^1(A(\varepsilon, 1))$. It follows that for large n the number of zeros of u_n is equal to the number of zeros of φ_k in the annulus $A(\varepsilon, 1)$, which is exactly k. Applying Proposition 4.1 we can assert that for large n there exists a zero $\varepsilon_n \in (0, \varepsilon]$ of u_n. Hence, multiplying the equation $-\Delta u_n = a_n u_n$ by u_n, integrating by parts in the ball B_{ε_n}, and applying Hölder inequality, we deduce

$$\int_{B_{\varepsilon_n}} |\nabla u_n|^2 = \int_{B_{\varepsilon_n}} a_n u_n^2 \leq \|a_n\|_{L^{N/2}(B_{\varepsilon_n})} \|u_n\|^2_{L^{2N/(N-2)}(B_{\varepsilon_n})}.$$

From the above it follows that

$$\|a_n\|_{L^{N/2}(B_{\varepsilon_n})} \geq \frac{\|\nabla u_n\|^2_{L^2(B_{\varepsilon_n})}}{\|u_n\|^2_{L^{2N/(N-2)}(B_{\varepsilon_n})}} \geq \inf_{u \in H_0^1(B_{\varepsilon_n})} \frac{\|\nabla u\|^2_{L^2(B_{\varepsilon_n})}}{\|u\|^2_{L^{2N/(N-2)}(B_{\varepsilon_n})}}.$$

From the change $v(x) = u(\varepsilon_n x)$, it is easily deduced that

$$\inf_{u \in H_0^1(B_{\varepsilon_n})} \frac{\|\nabla u\|^2_{L^2(B_{\varepsilon_n})}}{\|u\|^2_{L^{2N/(N-2)}(B_{\varepsilon_n})}} = \inf_{v \in H_0^1(B_1)} \frac{\|\nabla v\|^2_{L^2(B_1)}}{\|v\|^2_{L^{2N/(N-2)}(B_1)}} := C_N > 0.$$

From this we see, for fixed $\varepsilon \in (0, r_k)$ and large n, that

$$C_N \leq \|a_n\|_{L^{N/2}(B_{\varepsilon_n})} \leq \|a_n - \mu_k\|_{L^{N/2}(B_{\varepsilon_n})} + \|\mu_k\|_{L^{N/2}(B_{\varepsilon_n})}$$

$$\leq \|a_n - \mu_k\|_{L^{N/2}(B_1)} + \|\mu_k\|_{L^{N/2}(B_\varepsilon)}.$$

Taking limits when n tends to ∞ in this expression we deduce

$$C_N \leq \mu_k \left(\frac{\omega_N \varepsilon^N}{N} \right)^{2/N}.$$

Choosing $\varepsilon > 0$ sufficiently small we obtain a contradiction.

4.4.3 The Supercritical Case

In this subsection, we study the supercritical case, i.e., $p > \frac{N}{2}$, if $N \geq 2$. In all these cases we will prove that $\gamma_{p,k}$ is strictly positive and that it is attained. We begin by studying the case $p = \infty$.

Lemma 4.13. *Let $k \geq 0$, $N \geq 2$ and $p = \infty$. Then $\gamma_{\infty,k} = \mu_{k+1} - \mu_k$ is attained in the unique element $a_0 \equiv \mu_{k+1} \in \Gamma_k$.*

Proof. Clearly $a_0 \equiv \mu_{k+1} \in \Gamma_k$ satisfies $\|a_0 - \mu_k\|_{L^\infty(B_1)} = \mu_{k+1} - \mu_k$. Suppose, contrary to our claim, that there exists $\mu_{k+1} \not\equiv a \in \Gamma_k$ such that $\|a - \mu_k\|_{L^\infty(B_1)} \leq \mu_{k+1} - \mu_k$. Therefore $\mu_k \prec a \prec \mu_{k+1}$, a contradiction with the fact $a \in \Gamma_k$ (see [4, 10]).

Next we concentrate on the case $\frac{N}{2} < p < \infty$.

Lemma 4.14. *Let $k \geq 0$, $N \geq 2$, and $N/2 < p < \infty$. Then $\gamma_{p,k}$ is strictly positive and it is attained in a function $a_0 \in \Gamma_k$.*

Proof. Take a sequence $\{a_n\} \subset \Gamma_k$ such that $\|a_n - \mu_k\|_{L^p(B_1)} \to \gamma_{p,k}$. Take $\{u_n\} \subset H^1(B_1)$ such that u_n is a radial solution of (4.1), for $a = a_n$, with the normalization $\|u_n\|^2_{H^1(B_1)} = 1$. Therefore, we can suppose, up to a subsequence, that $u_n \rightharpoonup u_0$ in $H^1(B_1)$ and $u_n \to u_0$ in $L^{\frac{2p}{p-1}}(B_1)$ (since $p > N/2$, then $2 < \frac{2p}{p-1} < \frac{2N}{N-2}$, which is the critical Sobolev exponent). On the other hand, since $\{a_n\}$ is bounded in $L^p(B_1)$, and $1 \leq N/2 < p < \infty$, we can assume, up to a subsequence, that $a_n \rightharpoonup a_0$ in $L^p(B_1)$. Taking limits in Eq. (4.1), for $a = a_n$ and $u = u_n$, we obtain that u_0 is a solution of this equation for $a = a_0$. Note that $u_n \to u_0$ in $L^{\frac{2p}{p-1}}(B_1)$ and $a_n \rightharpoonup a_0$ in $L^p(B_1)$ yields $\lim \int_{B_1} |\nabla u_n|^2 = \lim \int_{B_1} a_n u_n^2 = \int_{B_1} a_0 u_0^2 = \int_{B_1} |\nabla u_0|^2$ and consequently $u_n \to u_0 \not\equiv 0$ in $H^1(B_1)$. Therefore, if $a_0 \not\equiv \mu_k$, then $a_0 \in \Gamma_k$ and $\|a_0 - \mu_k\|_p \leq \lim_{n\to\infty} \|a_n - \mu_k\|_p = \gamma_{p,k}$, and the lemma follows.

On the contrary, suppose by contradiction that $a_0 \equiv \mu_k$. Then $u_0 = \varphi_k$ for some nontrivial radial eigenfunction φ_k. First, we claim that $a_n \to \mu_k$ in $L^1(B_1)$. In effect, since $a_n \geq \mu_k$ and $a_n \rightharpoonup \mu_k$ in $L^p(B_1)$ we have that $\|a_n - \mu_k\|_{L^1(B_1)} = \int_\Omega (a_n - \mu_k) \to 0$, which is our claim. On the other hand, similar to the critical case, consider fixed $\varepsilon \in (0, r_k)$. Since $a_n \to \mu_k$ in $L^1(A(\varepsilon, 1))$ and $u_n \to u_0 = \varphi_k$

in $H_{\text{rad}}^1(A(\varepsilon, 1)) \subset C(A(\varepsilon, 1))$, we can assert that $a_n u_n \to \mu_k \varphi_k$ in $L^1(A(\varepsilon, 1))$. Thus $-\Delta u_n \to \mu_k \varphi_k$ in $L^1(A(\varepsilon, 1))$, which yields $u_n \to \varphi_k$ in $C^1(A(\varepsilon, 1))$. As the previous case, it follows that, for large n, the number of zeros of u_n is equal to the number of zeros of φ_k in the annulus $A(\varepsilon, 1)$, which is exactly k. Applying Proposition 4.1 we can assert that for large n there exists a zero $\varepsilon_n \in (0, \varepsilon]$ of u_n. Hence, multiplying the equation $-\Delta u_n = a_n u_n$ by u_n, integrating by parts in the ball B_{ε_n}, and applying Hölder inequality, we deduce

$$\int_{B_{\varepsilon_n}} |\nabla u_n|^2 = \int_{B_{\varepsilon_n}} a_n u_n^2 \le \|a_n\|_{L^p(B_{\varepsilon_n})} \|u_n\|_{L^{2p/(p-1)}(B_{\varepsilon_n})}^2.$$

From the above it is deduced that

$$\|a_n\|_{L^p(B_{\varepsilon_n})} \ge \frac{\|\nabla u_n\|_{L^2(B_{\varepsilon_n})}^2}{\|u_n\|_{L^{2p/(p-1)}(B_{\varepsilon_n})}^2} \ge \inf_{u \in H_0^1(B_{\varepsilon_n})} \frac{\|\nabla u\|_{L^2(B_{\varepsilon_n})}^2}{\|u\|_{L^{2p/(p-1)}(B_{\varepsilon_n})}^2}.$$

Again, from the change $v(x) = u(\varepsilon_n x)$, it follows easily that

$$\inf_{u \in H_0^1(B_{\varepsilon_n})} \frac{\|\nabla u\|_{L^2(B_{\varepsilon_n})}^2}{\|u\|_{L^{2p/(p-1)}(B_{\varepsilon_n})}^2} = \varepsilon_n^{N/p-2} \inf_{v \in H_0^1(B_1)} \frac{\|\nabla v\|_{L^2(B_1)}^2}{\|v\|_{L^{2p/(p-1)}(B_1)}^2} := \varepsilon_n^{N/p-2} C_{N,p} > 0.$$

From this, taking into account that $p > N/2$, we deduce, for fixed $\varepsilon \in (0, r_k)$ and large n, that

$$\varepsilon^{N/p-2} C_{N,p} \le \varepsilon_n^{N/p-2} C_{N,p} \le \|a_n\|_{L^{N/2}(B_{\varepsilon_n})} \le a_n\|_{L^{N/2}(B_1)}.$$

Finally, taking limits when ε tends to 0 in this expression we obtain that a_n is unbounded in $L^p(B_1)$, a contradiction.

References

1. Cañada, A., Montero, J.A., Villegas, S.: Lyapunov-type inequalities and Neumann boundary value problems at resonance. Math. Inequal. Appl. **8**, 459–475 (2005)
2. Cañada, A., Montero, J.A., Villegas, S.: Lyapunov inequalities for partial differential equations. J. Funct. Anal. **8**, 176–193 (2006)
3. Del Pino, M., Manásevich, R.F.: Global bifurcation from the eigenvalues of the p-Laplacian. J. Differ. Equ. **92**, 226–251 (1991)
4. Dolph, C.L.: Nonlinear integral equations of the Hammerstein type. Trans. Am. Math. Soc. **66**, 289–307 (1949)
5. Fuglede, B.: Continuous domain dependence of the eigenvalues of the Dirichlet Laplacian and related operators in Hilbert space. J. Funct. Anal. **167**, 183–200 (1999)
6. Hartman, P.: Ordinary Differential Equations. Wiley, New York (1964)

7. Huaizhong, W., Yong, L.: Neumann boundary value problems for second-order ordinary differential equations across resonance. SIAM J. Control Optim. **33**, 1312–1325 (1995)
8. Mawhin, J., Ward, J.R., Willem, M.: Variational methods and semilinear elliptic equations. Arch. Ration. Mech. Anal. **95**, 269–277 (1986)
9. Timoshin, S.: Lyapunov inequality for elliptic equations involving limiting nonlinearities. Proc. Jpn. Acad. Ser. A Math. Sci. **86**, 139–142 (2010)
10. Vidossich, G.: Existence and uniqueness results for boundary value problems from the comparison of eigenvalues. ICTP Preprint Archive, 1979015 (1979)

References

Chapter 5
Systems of Equations

Abstract This chapter is devoted to the study of L_p Lyapunov-type inequalities for linear systems of ordinary differential equations with different boundary conditions (which include the case of Neumann, Dirichlet, periodic, and antiperiodic boundary value problems) and for any constant $p \geq 1$. Elliptic problems are also considered. As in the scalar case, the results obtained in the linear case are combined with Schauder fixed point theorem to provide several results about the existence and uniqueness of solutions for resonant nonlinear systems. In addition, we study the stable boundedness of linear periodic conservative systems. The proof uses in a fundamental way the nontrivial relation (proved in Chap. 2) between the best Lyapunov constants and the minimum value of some especial constrained or unconstrained minimization problems (depending on the considered problems are resonant or nonresonant, respectively).

5.1 Motivation and the Meaning of Lyapunov Inequalities for Systems of Equations

Let us consider the scalar and linear Neumann b.v.p.

$$u''(x) + a(x)u(x) = 0, \ x \in (0, L), \ u'(0) = u'(L) = 0. \tag{5.1}$$

If function a satisfies

$$a \in L^\infty(0, L) \setminus \{0\}, \ \int_0^L a(x) \, dx \geq 0 \tag{5.2}$$

and

$$a^+ \prec \pi^2/L^2 \tag{5.3}$$

then (5.1) has only the trivial solution (remember that for $c, d \in L^1(0, L)$, we write $c \prec d$ if $c(x) \leq d(x)$ for a.e. $x \in [0, L]$ and $c(x) < d(x)$ on a set of positive measure).

In fact this is a trivial consequence of the L_∞ Lyapunov inequality showed in Chap. 2, Corollary 2.1.

© The Author(s) 2015
A. Cañada, S. Villegas, *A Variational Approach to Lyapunov Type Inequalities*,
SpringerBriefs in Mathematics, DOI 10.1007/978-3-319-25289-6_5

Several authors have generalized the assumptions (5.2)–(5.3) to vector differential equations of the form

$$u''(x) + A(x)u(x) = 0, \ x \in (0, L) \tag{5.4}$$

with different boundary conditions and where $A(\cdot)$ is a real and continuous $n \times n$ symmetric matrix valued function [1, 3, 14, 17, 26]. Also, some abstract versions for semilinear equations in Hilbert spaces and applications to elliptic problems and semilinear wave equations have been given in [2, 11, 20, 21].

In spite of its interest in the study of different questions such as stability theory, the calculation of lower bounds on eigenvalue problems, etc. [9, 12, 22, 28], the use of L_∞ Lyapunov inequalities in the study of nonlinear resonant problems only allows a weak interaction between the nonlinear term and the spectrum of the linear part. For example, using the L_∞ Lyapunov inequalities showed in [14] for the periodic boundary value problem (see also [1, 3]), it may be proved that if there exist real symmetric matrices P and Q with eigenvalues $p_1 \leq \cdots \leq p_n$ and $q_1 \leq \cdots \leq q_n$, respectively, such that

$$P \leq G''(u) \leq Q, \ \forall \, u \in \mathbf{R}^n \tag{5.5}$$

and such that

$$\bigcup_{i=1}^{n} [p_i, q_i] \cap \{k^2 : k \in \mathbf{N} \cup \{0\}\} = \emptyset \tag{5.6}$$

then, for each continuous and 2π-periodic function h, the periodic problem

$$u''(x) + G'(u(x)) = h(x), \ x \in (0, 2\pi), \ u(0) - u(2\pi) = u'(0) - u'(2\pi) = 0 \tag{5.7}$$

has a unique solution. Here $G : \mathbf{R}^n \to \mathbf{R}$ is a C^2-mapping and the relation $C \leq D$ between $n \times n$ matrices means that $D - C$ is positive semi-definite.

Let us explain in more detail the assertion *weak interaction between the nonlinear term and the spectrum of the linear part*. Using the variational characterization of the eigenvalues of a real symmetric matrix, it may be easily deduced that (5.5) and (5.6) imply that the eigenvalues $g_1(u) \leq \cdots \leq g_n(u)$ of the matrix $G''(u)$, satisfy

$$p_i \leq g_i(u) \leq q_i, \ \forall \, u \in \mathbf{R}^n. \tag{5.8}$$

Consequently each continuous function $g_i(u)$, $1 \leq i \leq n$, must fulfill

$$g_i(\mathbf{R}^n) \cap \{k^2 : k \in \mathbf{N} \cup \{0\}\} = \emptyset. \tag{5.9}$$

In this chapter we provide for each p, with $1 \leq p \leq \infty$, optimal necessary conditions for boundary value problem

$$u''(x) + A(x)u(x) = 0, \ x \in (0, L), \ u'(0) = u'(L) = 0 \tag{5.10}$$

to have nontrivial solutions (other boundary value problems like (5.7) can also be considered). For scalar equations, such optimal necessary conditions are the same as those obtained in Chaps. 2 and 4, where we used the corresponding best L_p Lyapunov constants (for example, Corollaries 2.1 and 2.2). In this monograph, this is the meaning of "Lyapunov inequalities for systems of equations."

The previously mentioned conditions are given in terms of the L^p norm of appropriate functions $b_{ii}(x)$, $1 \leq i \leq n$, related to $A(x)$ through the inequality $A(x) \leq B(x)$, $\forall x \in [0, L]$, where $B(x)$ is a diagonal matrix with entries given by $b_{ii}(x)$, $1 \leq i \leq n$. In particular, different L_{p_i} criteria for each $1 \leq i \leq n$ can be used and this confers a great generality to the shown results.

5.2 Neumann and Dirichlet Boundary Conditions

The next L_∞ *Theorem* is inspired from [14, 17], where the authors considered the periodic problem. The proof that we give here suggests the way for the corresponding L_p *Theorem*, when $1 \leq p < \infty$.

Theorem 5.1. *Let $A(\cdot)$ be a real $n \times n$ symmetric matrix valued function with elements defined and continuous on $[0, L]$. Suppose there exist diagonal matrix functions $P(x)$ and $Q(x)$ with continuous respective entries $\delta_{kk}(x)$, $1 \leq k \leq n$ and $\mu_{kk}(x)$, $1 \leq k \leq n$, and eigenvalues $\lambda_{p(k)}$, $1 \leq k \leq n$, of the eigenvalue problem*

$$u''(x) + \lambda u(x) = 0, \ x \in (0, L), \ u'(0) = u'(L) = 0 \tag{5.11}$$

such that

$$P(x) \leq A(x) \leq Q(x), \ \forall x \in [0, L] \tag{5.12}$$

and

$$\lambda_{p(k)} < \delta_{kk}(x) \leq \mu_{kk}(x) < \lambda_{p(k)+1}, \ \forall x \in [0, L], \ 1 \leq k \leq n. \tag{5.13}$$

Then there exists no nontrivial solution of (5.10).

Proof. If $u = (u_1, \cdots, u_n) \in (H^1(0, L))^n$ is a nontrivial solution of (5.10), then

$$\int_0^L \langle u'(x), v'(x) \rangle \, dx = \int_0^L \langle A(x)u(x), v(x) \rangle \, dx, \ \forall v \in (H^1(0, L))^n \tag{5.14}$$

where $\langle \cdot, \cdot \rangle$ is the usual scalar product in $\mathbf{R}^n$.

On the other hand, the eigenvalues of (5.11) are given by $\lambda_j = \frac{j^2 \pi^2}{L^2}$, where j is an arbitrary nonnegative integer number. If φ_j is the corresponding eigenfunction to λ_j, let us introduce the space $H = H_1 \times \cdots \times H_k \times \cdots \times H_n$, where for each $1 \leq k \leq n$, H_k is the span of the eigenfunctions $\varphi_0, \varphi_1, \cdots, \varphi_{p(k)}$.

It is trivial [6] that we can choose $\psi = (\psi_1, \cdots, \psi_n) \in H$ satisfying

$$u_k + \psi_k \in H_k^{\perp}, \; 1 \leq k \leq n. \tag{5.15}$$

The main ideas to get a contradiction with the fact that u is a nontrivial solution of (5.10) are the following two inequalities. The first one is a consequence of the variational characterization of the eigenvalues of (5.11). The second one is a trivial consequence of the definition of the subspace H_k.

$$\int_0^L ((u_k + \psi_k)'(x))^2 \, dx \geq \lambda_{p(k)+1} \int_0^L ((u_k + \psi_k)(x))^2 \, dx, \; 1 \leq k \leq n$$

$$\int_0^L ((\psi_k)'(x))^2 \, dx \leq \lambda_{p(k)} \int_0^L ((\psi_k)(x))^2 \, dx, \; 1 \leq k \leq n. \tag{5.16}$$

Now, from (5.14) we have

$$\int_0^L \langle (u + \psi)'(x), (u + \psi)'(x) \rangle \, dx = \int_0^L \langle A(x)(u + \psi)(x), (u + \psi)(x) \rangle \, dx$$

$$+ \int_0^L \langle \psi'(x), \psi'(x) \rangle \, dx$$

$$- \int_0^L \langle A(x)\psi(x), \psi(x) \rangle \, dx. \tag{5.17}$$

By using (5.12) and (5.13) we deduce

$$\int_0^L \langle \psi'(x), \psi'(x) \rangle \, dx - \int_0^L \langle A(x)\psi(x), \psi(x) \rangle \, dx$$

$$\leq \int_0^L \langle \psi'(x), \psi'(x) \rangle \, dx - \int_0^L \langle P(x)\psi(x), \psi(x) \rangle \, dx$$

$$= \sum_{k=1}^n \int_0^L [(\psi_k'(x))^2 - \delta_{kk}(x)(\psi_k(x))^2] \, dx$$

$$\leq \sum_{k=1}^n \int_0^L (\lambda_{p(k)} - \delta_{kk}(x))(\psi_k(x))^2 \, dx \leq 0.$$

Consequently,

$$\int_0^L \langle (u+\psi)'(x), (u+\psi)'(x) \rangle \, dx \leq \int_0^L \langle A(x)(u+\psi)(x), (u+\psi)(x) \rangle \, dx. \tag{5.18}$$

Also, from (5.12), (5.15), (5.16), and (5.18) we obtain

$$\sum_{k=1}^{n} \int_{0}^{L} \lambda_{p(k)+1}(u_k + \psi_k)^2(x)\, dx \leq \sum_{k=1}^{n} \int_{0}^{L} (u_k + \psi_k)'^2(x)\, dx$$

$$= \int_{0}^{L} \langle (u + \psi)'(x), (u + \psi)'(x) \rangle\, dx$$

$$\leq \int_{0}^{L} \langle A(x)(u + \psi)(x), (u + \psi)(x) \rangle\, dx$$

$$\leq \int_{0}^{L} \langle Q(x)(u + \psi)(x), (u + \psi)(x) \rangle\, dx$$

$$= \sum_{k=1}^{n} \int_{0}^{L} \mu_{kk}(x)(u_k + \psi_k)^2(x)\, dx. \qquad (5.19)$$

It follows, again from (5.13), that

$$u + \psi \equiv 0. \qquad (5.20)$$

But if $u + \psi \equiv 0$, then $u = \phi = (\phi_1, \cdots, \phi_n)$ for some nontrivial $\phi \in H$. Therefore,

$$\sum_{k=1}^{n} \int_{0}^{L} \lambda_{p(k)}(\phi_k)^2(x)\, dx \geq \sum_{k=1}^{n} \int_{0}^{L} (\phi_k)'^2(x)\, dx$$

$$= \int_{0}^{L} < \phi'(x), \phi'(x)\, dx = \int_{0}^{L} \langle A(x)\phi(x), \phi(x) \rangle\, dx$$

$$\geq \int_{0}^{L} \langle P(x)\phi(x), \phi(x) \rangle\, dx = \sum_{k=1}^{n} \int_{0}^{L} \delta_{kk}(x)(\phi_k)^2(x)\, dx.$$

$$(5.21)$$

Now, (5.13) implies that $u_k = \phi_k \equiv 0$, $1 \leq k \leq n$, which is a contradiction with the fact that u is nontrivial.

Next Lemma is an essential step, prior to the consideration of the case $1 \leq p < \infty$ for systems of equations. Really, the proof is contained in Theorems 2.1, 2.2, 2.3 and Corollary 2.2, but it is again convenient to highlight these facts here.

Lemma 5.1. *If* $1 \leq p \leq \infty$ *is a given number, let us define the subset* X_p *of* $H^1(0, L)$ *and the functional* I_p *as*

$$X_1 = \left\{ v \in H^1(0,L) : \max_{x \in [0,L]} v(x) + \min_{x \in [0,L]} v(x) = 0 \right\}$$

$$I_1 : X_1 \setminus \{0\} \to \mathbf{R}, \ I_1(v) = \frac{\int_0^L v'^2}{\|v\|_\infty^2}$$

$$X_p = \left\{ v \in H^1(0,L) : \int_0^L |v|^{\frac{2}{p-1}} v = 0 \right\}, \ \text{if} \ 1 < p < \infty$$

$$I_p : X_p \setminus \{0\} \to \mathbf{R}, \ I_p(v) = \frac{\int_0^L v'^2}{\left(\int_0^L |v|^{\frac{2p}{p-1}} \right)^{\frac{p-1}{p}}}, \ \text{if} \ 1 < p < \infty \tag{5.22}$$

$$X_\infty = \left\{ v \in H^1(0,L) : \int_0^L v = 0 \right\}$$

$$I_\infty : X_\infty \setminus \{0\} \to \mathbf{R}, \ I_\infty(v) = \frac{\int_0^L v'^2}{\int_0^L v^2}.$$

Then, β_p, the best L_p Lyapunov constant defined in (2.5), satisfies

$$\beta_p \equiv \min_{X_p \setminus \{0\}} I_p, \ 1 \leq p \leq \infty \tag{5.23}$$

and consequently

1. $\displaystyle\int_0^L v'^2 \geq \beta_1 \|v\|_\infty^2, \ \forall \ v \in H^1(0,L) \cap X_1$

2. $\displaystyle\int_0^L v'^2 \geq \beta_p \left(\int_0^L |v|^{\frac{2p}{p-1}} \right)^{\frac{p-1}{p}}, \ \forall \ v \in H^1(0,L) \cap X_p, \ 1 < p < \infty$

3. $\displaystyle\int_0^L v'^2 \geq \beta_\infty \int_0^L v^2, \ \forall \ v \in H^1(0,L) \cap X_\infty$

Moreover, and as a consequence of the definition of β_p in (2.5), if for some $p \in [1,\infty]$, function a satisfies (5.2) and $\|a^+\|_p < \beta_p$, then (5.1) has only the trivial solution.

We return to system (5.10). From now on, we assume that the matrix function $A(\cdot) \in \Lambda$ where Λ is defined as

The set of real $n \times n$ symmetric matrix valued function $A(\cdot)$, with continuous element functions $a_{ij}(x)$, $1 \leq i,j \leq n$, $x \in [0,L]$, such that (5.10) has not nontrivial constant solutions and

[Λ]

$$\int_0^L \langle A(x)k, k \rangle dx \geq 0, \ \forall \ k \in \mathbf{R}^n.$$

In particular, if $A(\cdot)$ is a real, continuous $n \times n$ symmetric matrix valued function such that $A(x)$ is positive semi-definite, $\forall \ x \in [0,L]$ and $\det A(x) \neq 0$ for some $x \in [0,L]$, then $A(\cdot) \in \Lambda$ (here $\det A(x)$ means the determinant of the matrix $A(x)$). If $n = 1$, these conditions are slightly more restrictive than (5.2).

The main result of this section is the following (see [6] for a slightly more general result).

Theorem 5.2. *Let $A(\cdot) \in \Lambda$ be such that there exist a diagonal matrix $B(x)$ with continuous entries $b_{ii}(x)$, and $p_i \in [1, \infty]$, $1 \leq i \leq n$, satisfying*

$$A(x) \leq B(x), \ \forall \ x \in [0, L]$$

$$\|b_{ii}^+\|_{p_i} < \beta_{p_i}, \ 1 \leq i \leq n. \tag{5.24}$$

Then, there exists no nontrivial solution of the vector boundary value problem (5.10).

Proof. If $u \in H^1(0, L)$ is any nontrivial solution of (5.10), we have

$$\int_0^L \langle eu'(x), v'(x) \rangle = \int_0^L \langle A(x)u(x), v(x) \rangle, \ \forall \ v \in H^1(0, L).$$

In particular, we have

$$\int_0^L \langle u'(x), u'(x) \rangle = \int_0^L \langle A(x)u(x), u(x) \rangle$$

$$\int_0^L \langle A(x)u(x), k \rangle = \int_0^L \langle A(x)k, u(x) \rangle = 0, \ \forall \ k \in \mathbf{R}^n. \tag{5.25}$$

Therefore, for each $k \in \mathbf{R}^n$, we have

$$\int_0^L \langle (u(x) + k)', (u(x) + k)' \rangle = \int_0^L \langle u'(x), u'(x) \rangle$$

$$= \int_0^L \langle A(x)u(x), u(x) \rangle \leq \int_0^L \langle A(x)u(x), u(x) \rangle$$

$$+ \int_0^L \langle A(x)u(x), k \rangle + \int_0^L \langle A(x)k, u(x) \rangle$$

$$+ \int_0^L \langle A(x)k, k \rangle = \int_0^L \langle A(x)(u(x) + k), u(x) + k \rangle$$

$$\leq \int_0^L \langle B(x)(u(x) + k), u(x) + k \rangle.$$

If $u = (u_i)$, then for each $i, 1 \leq i \leq n$, we choose $k_i \in \mathbf{R}$ satisfying $u_i + k_i \in X_{p_i}$, the set defined in Lemma 5.1. By using previous inequality, Lemma 5.1 and Hölder inequality, we obtain

$$\sum_{i=1}^n \beta_{p_i} \| (u_i + k_i)^2 \|_{\frac{p_i}{p_i-1}} \leq \sum_{i=1}^n \int_0^L (u_i(x) + k_i)'^2$$

$$\leq \sum_{i=1}^n \int_0^L b_{ii}^+(x)(u_i(x) + k_i)^2$$

$$\leq \sum_{i=1}^n \| b_{ii}^+ \|_{p_i} \| (u_i + k_i)^2 \|_{\frac{p_i}{p_i-1}}, \qquad (5.26)$$

where

$$\frac{p_i}{p_i-1} = \infty, \ \ \text{if } p_i = 1$$

$$\frac{p_i}{p_i-1} = 1, \ \ \text{if } p_i = \infty.$$

Therefore from (5.24) we have

$$\sum_{i=1}^n (\beta_{p_i} - \| b_{ii}^+ \|_{p_i}) \| (u_i + k_i)^2 \|_{\frac{p_i}{p_i-1}} \leq 0. \qquad (5.27)$$

On the other hand, since u is a nontrivial function, $u + k$ is also a nontrivial function. Indeed, if $u + k$ is identically zero, we deduce that (5.10) has the nontrivial and constant solution $-k$ which is a contradiction with the hypothesis $A(\cdot) \in \Lambda$.

Now, if $u + k$ is nontrivial, some component, say, $u_j + k_j$ is nontrivial. Then, $(\beta_{p_j} - \| b_{jj}^+ \|_{p_j}) \| (u_j + k_j)^2 \|_{\frac{p_j}{p_j-1}}$ is strictly positive and from (5.24), all the other summands in (5.27) are nonnegative. This is a contradiction.

Remark 5.1. Previous theorem is optimal in the following sense. For any given positive numbers γ_i, $1 \leq i \leq n$, such that at least one of them, say γ_j, satisfies

$$\gamma_j > \beta_{p_j}, \ \text{for some } p_j \in [1, \infty] \qquad (5.28)$$

there exists a diagonal $n \times n$ matrix $A(\cdot) \in \Lambda$ with continuous entries $a_{ii}(x)$, $1 \leq i \leq n$, satisfying $\|a_{ii}^+\|_{p_i} < \gamma_i$, $1 \leq i \leq n$ and such that the boundary value problem (5.10) has nontrivial solutions. To see this, if γ_j satisfies (5.28), then by the definition of β_{p_j} in (2.5), there exists some continuous function $a(x)$, not identically zero, with $\int_0^L a(x)\,dx \geq 0$, and $\|a^+\|_{p_j} < \gamma_j$, such that the scalar problem

$$w''(x) + a(x)w(x) = 0, \ x \in (0, L), \ w'(0) = w'(L) = 0$$

has nontrivial solutions. Then, to get our purpose, it is sufficient to take $a_{jj}(x) = a(x)$ and $a_{ii}(x) = \delta \in \mathbf{R}^+$, if $i \neq j$, with δ sufficiently small.

As an application of Theorem 5.2 we have the following corollary.

Corollary 5.1. *Let $A(\cdot) \in \Lambda$ and, for each $x \in [0, L]$, let us denote by $\rho(x)$ the spectral radius of the matrix $A(x)$. If the function $\rho(\cdot)$ satisfies*

$$\|\rho^+\|_p < \beta_p$$

for some $p \in [1, \infty]$, then there exists no nontrivial solution of (5.10).

Proof. It is trivial, taking into account the previous theorem and the inequality

$$A(x) \leq \rho(x)I_n, \ \forall \, x \in [0, L], \tag{5.29}$$

where I_n is the $n \times n$ identity matrix.

Remark 5.2. With regard to Theorem 5.2, the authors introduced in [14, 17] similar conditions for periodic problems and $p_i = \infty$, $1 \leq i \leq n$. Our variational method of proof, where we strongly use the minimization problems considered in Lemma 5.1, does possible the consideration of the cases $p \in [1, \infty)$. Let us emphasize that if $p \in [1, \infty)$ in the previous Corollary, then the function $\rho(x)$ may cross an arbitrary number of eigenvalues of the problem (5.11).

Remark 5.3. If $A(\cdot)$ satisfies

[H] $A(x)$, $x \in [0, L]$ *is a continuous and positive semi-definite matrix function such that* $\det A(x) \neq 0$ *for some $x \in [0, L]$*

and

$$\int_0^L \text{trace}\, A(x)\,dx < \beta_1 \tag{5.30}$$

then there exists no nontrivial solution of (5.10). In fact, taking into account that for each $x \in [0, L]$, $\rho(x)$ is an eigenvalue of the matrix $A(x)$ and that in this case all the eigenvalues of $A(x)$, $\lambda_1(x), \cdots, \lambda_n(x)$, are nonnegative, we have $\rho(x) \leq \sum_{i=1}^n \lambda_i(x) = \text{trace}\, A(x)$ (see [15] for this last relation). Therefore, from (5.30) we obtain

$$\|\rho^+\|_1 = \int_0^L \rho(x)\,dx < \beta_1. \tag{5.31}$$

Previous remark shows that if we want to have a criterion implying that (5.10) has only the trivial solution, then (5.31) is better than (5.30). Similar results, contained in [23–25], have been obtained for Dirichlet boundary conditions.

Remark 5.4. Similar results to those given by Theorem 5.2 can be obtained for Dirichlet boundary conditions, but in this case, the hypothesis [H] is not necessary. However, for Neumann boundary conditions, a restriction like [H] is natural (see Remarks 4 and 5 in [4]).

5.3 Periodic and Antiperiodic Boundary Conditions

A similar study may be performed for other boundary conditions. Because of the applications shown in the next two sections to nonlinear resonant periodic systems and to the study of stability of linear periodic systems, respectively, we highlight the case of periodic and antiperiodic boundary conditions. The proof of the next two theorems is an easy consequence of the results shown in Sect. 2.3 (more precisely, Theorems 2.5 and 2.6), where the variational characterization of the best L_p Lyapunov constant for these two types of boundary value problems was proved. See [6] for further details.

For the periodic boundary value problem

$$u''(t) + a(t)u(t) = 0,\ t \in (0,T),\ u(0) - u(T) = u'(0) - u'(T) = 0 \tag{5.32}$$

and for the antiperiodic boundary value problem

$$u''(t) + a(t)u(t) = 0,\ t \in (0,T),\ u(0) + u(T) = u'(0) + u'(T) = 0 \tag{5.33}$$

we assume that $a \in L_T(\mathbf{R}, \mathbf{R})$, the set of T-periodic functions $a : \mathbf{R} \to \mathbf{R}$ such that $a|_{[0,T]} \in L^1(0,T)$.

Remember (Sect. 2.3) that if we define the sets

$$\Lambda^{\mathrm{per}} = \left\{ a \in L_T(\mathbf{R},\mathbf{R}) \setminus \{0\} : \int_0^T a(t)dt \geq 0 \text{ and } (5.32) \text{ has nontrivial solutions} \right\}$$
$$\tag{5.34}$$

$$\Lambda^{\mathrm{ant}} = \{ a \in L_T(\mathbf{R},\mathbf{R}) : (5.33) \text{ has nontrivial solutions} \} \tag{5.35}$$

then, for each p with $1 \leq p \leq \infty$, we can define, respectively, the L^p Lyapunov constants β_p^{per} and β_p^{ant} as the real numbers

$$\beta_p^{\mathrm{per}} \equiv \inf_{a \in \Lambda^{\mathrm{per}} \cap L^p(0,T)} \|a^+\|_p, \quad \beta_p^{\mathrm{ant}} \equiv \inf_{a \in \Lambda^{\mathrm{ant}} \cap L^p(0,T)} \|a^+\|_p. \tag{5.36}$$

Theorem 5.3. *If* $1 \leq p \leq \infty$ *is a given number, let us define the sets* X_p^{per} *and the functionals* $I_p^{per} : X_p^{per} \setminus \{0\} \to \mathbf{R}$ *as*

$$X_1^{per} = \left\{ v \in H^1(0,T) : v(0) - v(T) = 0, \max_{t \in [0,T]} v(t) + \min_{t \in [0,T]} v(t) = 0 \right\},$$

$$X_p^{per} = \left\{ v \in H^1(0,T) : v(0) - v(T) = 0, \int_0^T |v|^{\frac{2}{p-1}} v = 0 \right\}, \ \text{if } 1 < p < \infty,$$

$$X_\infty^{per} = \left\{ v \in H^1(0,T) : v(0) - v(T) = 0, \int_0^T v = 0 \right\},$$

$$I_1^{per}(v) = \frac{\displaystyle\int_0^T v'^2}{\|v\|_\infty^2}, \quad I_p^{per}(v) = \frac{\displaystyle\int_0^T v'^2}{\left(\displaystyle\int_0^T |v|^{\frac{2p}{p-1}}\right)^{\frac{p-1}{p}}}, \ \text{if } 1 < p < \infty,$$

$$I_\infty^{per}(v) = \frac{\displaystyle\int_0^T v'^2}{\displaystyle\int_0^T v^2}. \tag{5.37}$$

Then, β_p^{per}, *the best* L_p *Lyapunov constant for the periodic problem, defined in (5.36), satisfies*

$$\beta_p^{per} \equiv \min_{X_p^{per} \setminus \{0\}} I_p^{per}, \ 1 \leq p \leq \infty \tag{5.38}$$

and consequently

1. $\displaystyle\int_0^T v'^2 \geq \beta_1^{per} \|v\|_\infty^2, \ \forall \, v \in H^1(0,T) \cap X_1^{per}$

2. $\displaystyle\int_0^T v'^2 \geq \beta_p^{per} \left(\int_0^T |v|^{\frac{2p}{p-1}}\right)^{\frac{p-1}{p}}, \ 1 < p < \infty, \ \forall \, v \in H^1(0,T) \cap X_p^{per}$

3. $\displaystyle\int_0^T v'^2 \geq \beta_\infty^{per} \int_0^T v^2, \ \forall \, v \in H^1(0,T) \cap X_\infty^{per}$

Moreover, if for some $p \in [1,\infty]$, *function* a *satisfies*

$$a \in L_T(\mathbf{R},\mathbf{R}) \setminus \{0\}, \ \int_0^T a(t) \, dt \geq 0$$

and $\|a^+\|_p < \beta_p^{per}$, *then (5.32) has only the trivial solution.*

Theorem 5.4. *If $1 \le p \le \infty$ is a given number, let us define the sets X_p^{ant} and the functional $I_p^{ant} : X_p^{ant} \setminus \{0\} \to \mathbf{R}$, as*

$$X_p^{ant} = \{v \in H^1(0,T) : v(0) + v(T) = 0\}, \ 1 \le p \le \infty$$

$$I_1^{ant}(v) = \frac{\int_0^T v'^2}{\|v\|_\infty^2}, \ I_p^{ant}(v) = \frac{\int_0^T v'^2}{\left(\int_0^T |v|^{\frac{2p}{p-1}}\right)^{\frac{p-1}{p}}}, \ if \ 1 < p < \infty,$$

$$I_\infty^{ant}(v) = \frac{\int_0^T v'^2}{\int_0^T v^2}. \tag{5.39}$$

Then, β_p^{ant}, the best L_p Lyapunov constant for the antiperiodic problem, defined in (5.36), satisfies

$$\beta_p^{ant} \equiv \min_{X_p^{ant}\setminus\{0\}} I_p^{ant}, \ 1 \le p \le \infty \tag{5.40}$$

and consequently

1. $\int_0^T v'^2 \ge \beta_1^{ant}\|v\|_\infty^2, \ \forall \ v \in H^1(0,T) \cap X_1^{ant}$

2. $\int_0^T v'^2 \ge \beta_p^{ant} \left(\int_0^T |v|^{\frac{2p}{p-1}}\right)^{\frac{p-1}{p}}, \ 1 < p < \infty, \ \forall \ v \in H^1(0,T) \cap X_p^{ant}$

3. $\int_0^T v'^2 \ge \beta_\infty^{ant} \int_0^T v^2, \ \forall \ v \in H^1(0,T) \cap X_\infty^{ant}$

Moreover, if for some $p \in [1,\infty]$, function $a \in L_T(\mathbf{R},\mathbf{R})$ satisfies $\|a^+\|_p < \beta_p^{ant}$, then (5.33) has only the trivial solution.

5.4 Nonlinear Resonant Problems

As we have shown in the previous chapters, an important application of Lyapunov inequalities is in the study of nonlinear resonant problems. In this section we deal with nonlinear systems of equations, a much more complicated topic than the case of scalar equations. We deal with periodic conditions, but obviously other boundary problems may be considered.

Let us begin motivating the problem. If $G : \mathbf{R}^n \to \mathbf{R}$ is a C^2-mapping and A and B are real symmetric $n \times n$ matrices with respective eigenvalues $a_1 \le \dots \le a_n$ and $b_1 \le \dots \le b_n$ satisfying

$$A \leq G''(u) \leq B, \ \forall \, u \in \mathbf{R}^n$$

$$0 < a_i \leq b_i < \frac{4\pi^2}{T^2}, \ 1 \leq i \leq n \tag{5.41}$$

then, for each continuous and T-periodic function $h : \mathbf{R} \to \mathbf{R}^n$, the periodic problem

$$u''(t) + G'(u(t)) = h(t), \ t \in (0, T), \ u(0) - u(T) = u'(0) - u'(T) = 0 \tag{5.42}$$

has a unique solution [1, 3, 14]. As it was mentioned in the first section of this chapter, this last result is also true under more general restrictions than (5.41) which involves higher eigenvalues of the periodic problem

$$u''(t) + \lambda u(t) = 0, \ t \in (0, T), \ u(0) - u(T) = u'(0) - u'(T) = 0. \tag{5.43}$$

The hypothesis (5.41) allows only a weak interaction between $G''(u)$ and the spectrum of the linear part (5.43) in the following sense: by using the variational characterization of the eigenvalues of a real symmetric matrix, it may be easily deduced that (5.41) implies that the eigenvalues $g_1(u) \leq \cdots \leq g_n(u)$ of the matrix $G''(u)$ satisfy

$$0 < a_i \leq g_i(u) \leq b_i < \frac{4\pi^2}{T^2}, \ \forall \, u \in \mathbf{R}^n, \ 1 \leq i \leq n. \tag{5.44}$$

In fact, it can be affirmed that the L^∞ restriction (5.41) is a nonresonant hypothesis.

In this section we provide for each p, with $1 \leq p \leq \infty$, L^p restrictions for boundary value problem (5.42) to have a unique solution. These are optimal in the sense shown in Remark 5.5 below. They are given in terms of the L^p norm of appropriate functions $b_{ii}(t)$, $1 \leq i \leq n$, related to (5.42) through the inequality $A(t) \leq G''(u) \leq B(t)$, $\forall \, t \in [0, T]$, where $B(t)$ is a diagonal matrix with entries given by $b_{ii}(t)$, $1 \leq i \leq n$ and $A(t)$ is a convenient symmetric matrix which belongs to Λ (this avoids the resonance at the eigenvalue 0). Since our conditions are given in terms of L^p norms, we allow to the eigenvalues $g_i(u)$, $1 \leq i \leq n$, to cross an arbitrary number of eigenvalues of (5.43), as long as certain L^p norms are controlled. The next theorem is an example of it. It can be applied to the study of systems of equations of the type (5.46) below, which models the Newtonian equation of motion of a mechanical system subject to conservative internal forces and periodic external forces.

Theorem 5.5. *Let $G : \mathbf{R} \times \mathbf{R}^n \to \mathbf{R}$, $(t, u) \to G(t, u)$, be a continuous function, T-periodic with respect to the variable t and satisfying:*

1. *For each $t \in \mathbf{R}$, the mapping $u \to G(t, u)$ is of class $C^2(\mathbf{R}^n, \mathbf{R})$*
2. *There exist continuous and T-periodic matrix functions $A(\cdot)$, $B(\cdot)$, with $A(t)$ symmetric, $B(t)$ diagonal, with entries $b_{ii}(t)$, and $p_i \in [1, \infty]$ $1 \leq i \leq n$, such that*

$$A(t) \le G_{uu}(t, u) \le B(t), \ \forall (t, u) \in \mathbf{R}^{n+1}$$

$$\int_0^T \ < A(t)k, k > \ dt > 0, \ \forall k \in \mathbf{R}^n \setminus \{0\}$$

$$\|b_{ii}^+\|_{p_i} < \beta_{p_i}^{per}, \ 1 \le i \le n$$

(5.45)

Then the boundary value problem

$$u''(t) + G_u(t, u(t)) = 0, \ t \in \mathbf{R}, \ u(0) - u(T) = u'(0) - u'(T) = 0 \qquad (5.46)$$

has a unique solution.

Proof. As in Theorem 2.4, it is based on two steps: first, we prove the uniqueness property. This suggests the way to prove existence of solutions [5].

1. Uniqueness of solutions of (5.46).

Let us denote by $H_T^1(0, T)$ the subset of T-periodic functions of the Sobolev space $H^1(0, T)$. Then, if $v \in (H_T^1(0, T))^n$ and $w \in (H_T^1(0, T))^n$ are two solutions of (5.46), the function $u = v - w$ is a solution of the problem

$$u''(t) + C(t)u(t) = 0, \ t \in \mathbf{R}, \ u(0) - u(T) = u'(0) - u'(T) = 0, \qquad (5.47)$$

where

$$C(t) = \int_0^1 G_{uu}(t, w(t) + \theta u(t)) \, d\theta$$

(See [16], p. 103, for the mean value theorem for the vectorial function $G_u(t, u)$.)
From (5.45),

$$A(t) \le C(t) \le B(t), \ \forall \, t \in \mathbf{R} \qquad (5.48)$$

and

$$\int_0^T \langle u'(t), z'(t) \rangle \ = \int_0^T \langle C(t)u(t), z(t) \rangle, \ \forall \, z \in (H_T^1(0, T))^n.$$

In particular, we have

$$\int_0^T \langle u'(t), u'(t) \rangle \ = \int_0^T \langle C(t)u(t), u(t) \rangle$$

$$\int_0^T \langle C(t)u(t), k \rangle \ = \int_0^T \langle C(t)k, u(t) \rangle \ = 0, \ \forall \, k \in \mathbf{R}^n. \qquad (5.49)$$

Therefore, by using again (5.45), for each $k \in \mathbf{R}^n$ we have

$$
\begin{aligned}
\int_0^T \langle (u(t) + k)', (u(t) + k)' \rangle &= \int_0^T \langle u'(t), u'(t) \rangle \\
&= \int_0^T \langle C(t)u(t), u(t) \rangle \leq \int_0^T \langle C(t)u(t), u(t) \rangle \\
&\quad + \int_0^T \langle C(t)u(t), k \rangle + \int_0^T \langle C(t)k, u(t) \rangle \\
&\quad + \int_0^T < C(t)k, k > \\
&= \int_0^T \langle C(t)(u(t) + k), u(t) + k \rangle \\
&\leq \int_0^T \langle B(t)(u(t) + k), u(t) + k \rangle.
\end{aligned}
$$

If $u = (u_i)$, then for each i, $1 \leq i \leq n$, we choose $k_i \in \mathbf{R}$ such that $u_i + k_i \in X_{p_i}^{\mathrm{per}}$, the set defined in Lemma 5.3. By using previous inequality, Lemma 5.3 and Hölder inequality, we obtain

$$
\begin{aligned}
\sum_{i=1}^n \beta_{p_i}^{\mathrm{per}} \| (u_i + k_i)^2 \|_{\frac{p_i}{p_i - 1}} &\leq \sum_{i=1}^n \int_0^T (u_i(t) + k_i)'^2 \\
&\leq \sum_{i=1}^n \int_0^T b_{ii}^+(t)(u_i(t) + k_i)^2 \leq \sum_{i=1}^n \| b_{ii}^+ \|_{p_i} \| (u_i + k_i)^2 \|_{\frac{p_i}{p_i - 1}},
\end{aligned}
\tag{5.50}
$$

where

$$
\frac{p_i}{p_i - 1} = \infty, \quad \text{if } p_i = 1
$$

$$
\frac{p_i}{p_i - 1} = 1, \quad \text{if } p_i = \infty.
$$

Therefore we have

$$
\sum_{i=1}^n (\beta_{p_i}^{\mathrm{per}} - \| b_{ii}^+ \|_{p_i}) \| (u_i + k_i)^2 \|_{\frac{p_i}{p_i - 1}} \leq 0.
\tag{5.51}
$$

But (5.51) implies that, necessarily, $u \equiv 0$ (and as a consequence, $v \equiv w$). To see this, if u is not identically zero, then the function $u + k$ is also not identically zero. In fact, if $u + k \equiv 0$, we deduce that (5.47) has the nontrivial and constant solution $-k$ which imply

$$0 = \int_0^T \langle C(t)k, k \rangle \, dt \geq \int_0^T \langle A(t)k, k \rangle \, dt.$$

This is a contradiction with (5.45).

Now, since $u + k$ is a nontrivial function, some component, say $u_j + k_j$, is nontrivial. Then $(\beta_{p_j}^{\text{per}} - \|b_{jj}^+\|_{p_j})\|(u_j + k_j)^2\|_{\frac{p_j}{p_j-1}}$ is strictly positive and from (5.45), all the other summands in (5.51) are nonnegative. This is a contradiction with (5.51).

2. Existence of solutions of (5.46).

As in the scalar case (Theorem 2.4), the proof of the existence of solutions of (5.46), is a combination of two main ideas: the use of some previous results on the existence and uniqueness of solutions of linear systems, together with the Schauder fixed point theorem.

First, we write (5.46) in the equivalent form

$$\left. \begin{array}{c} u''(t) + D(t, u(t))u(t) + G_u(t, 0) = 0, \ t \in \mathbf{R} \\ u(0) - u(T) = u'(0) - u'(T) = 0 \end{array} \right\}, \tag{5.52}$$

where the function $D : \mathbf{R} \times \mathbf{R}^n \to \mathscr{M}(\mathbf{R})$ is defined by $D(t, z) = \int_0^1 G_{uu}(t, \theta z) \, d\theta$. Here $\mathscr{M}(\mathbf{R})$ denotes the set of real $n \times n$ matrices.

If $C_T(\mathbf{R}, \mathbf{R})$ is the set of real T-periodic and continuous functions defined in $\mathbf{R}$, let us denote $X = (C_T(\mathbf{R}, \mathbf{R}))^n$ with the uniform norm, i.e., if $y(\cdot) = (y^1(\cdot), \cdots, y^n(\cdot)) \in X$, then $\|y\|_X = \sum_{k=1}^n \|y^k(\cdot)\|_\infty$.

Since

$$A(t) \leq D(t, z) \leq B(t), \ \forall \ (t, z) \in \mathbf{R} \times \mathbf{R}^n \tag{5.53}$$

we can define the operator $H : X \to X$, by $Hy = u_y$, being u_y the unique solution of the linear problem

$$\left. \begin{array}{c} u''(t) + D(t, y(t))u(t) = -G_u(t, 0), \ t \in \mathbf{R} \\ u(0) - u(T) = u'(0) - u'(T) = 0 \end{array} \right\}. \tag{5.54}$$

Let us remark that (5.54) is a nonhomogeneous linear problem such that the corresponding homogeneous one has only the trivial solution (as in the previous step on uniqueness).

We will show that H is completely continuous and that $H(X)$ is bounded. The Schauder's fixed point theorem [10] provides a fixed point for H which is a solution of (5.46).

The fact that H is completely continuous is a consequence of the Arzela–Ascoli Theorem [7]. Also, $H(X)$ is bounded. Suppose, contrary to our claim, that $H(X)$ is not bounded. In this case, there would exist a sequence $\{y_n\} \subset X$ such that

$\|u_{y_n}\|_X \to \infty$. From (5.53), and passing to a subsequence if necessary, we may assume that, for each $1 \leq i,j \leq n$, the sequence of functions $\{D_{ij}(\cdot, y_n(\cdot))\}$ is weakly convergent in $L^p(\Omega)$ to a function $E_{ij}(\cdot)$ and such that if $E(t) = (E_{ij}(t))$, then $A(t) \leq E(t) \leq B(t)$, a.e. in $\mathbf{R}$, ([18], p. 157).

If $z_n \equiv \dfrac{u_{y_n}}{\|u_{y_n}\|_X}$, passing to a subsequence if necessary, we may assume that $z_n \to z_0$ strongly in X, where z_0 is a nonzero vectorial function satisfying

$$\left.\begin{array}{c} z_0''(t) + E(t)z_0(t) = 0, \ t \in \mathbf{R} \\ z_0(0) - z_0(T) = z_0'(0) - z_0'(T) = 0 \end{array}\right\} . \qquad (5.55)$$

But, $A(t) \leq E(t) \leq B(t)$, $\forall\, t \in \mathbf{R}$ and, as in the first step on uniqueness, this implies that the unique solution of (5.55) is the trivial one. This is a contradiction with the fact that $\|z_0\|_X = 1$ and, as a consequence, $H(X)$ is bounded.

Remark 5.5. It can be shown that previous theorem is optimal in the same sense as the Remark 5.1 (see [5]).

Example 5.1. Here we show an example where it is allowed to the eigenvalues of the matrix $G_{uu}(t, u)$, in the boundary value problem (5.46), to cross an arbitrary number of eigenvalues of (5.43).

To begin with the example, remember a known result. Let $H : \mathbf{R}^n \to \mathbf{R}$, $u \to H(u)$ be a given function of class $C^2(\mathbf{R}^n, \mathbf{R})$ such that

1. There exist a real constant symmetric $n \times n$ matrix A and a real constant diagonal matrix B, with respective eigenvalues

$$a_1 \leq a_2 \leq \ldots \leq a_n$$
$$b_1 \leq b_2 \leq \ldots \leq b_n$$

satisfying $A \leq H_{uu}(u) \leq B$, $\forall\, u \in \mathbf{R}^n$
2. $0 < a_k \leq b_k < 1$, $1 \leq k \leq n$

Then for each continuous and 2π-periodic function $h : \mathbf{R} \to \mathbf{R}^n$, the boundary value problem

$$u''(t) + H_u(u(t)) = h(t), \ t \in \mathbf{R}, \ u(0) - u(2\pi) = u'(0) - u'(2\pi) = 0 \qquad (5.56)$$

has a unique solution [1, 14]. Let us remark that since $T = 2\pi$, then the numbers 0 and 1 are the first two eigenvalues of the eigenvalue problem (5.43).

Now, to get our purpose, we slightly modify the previous example in the following way: if $m : \mathbf{R} \to \mathbf{R}$, is a given continuous and 2π-periodic function such that for some $p_i \in [1, \infty]$, $1 \leq i \leq n$,

$$m(t) \geq 0, \ \forall\, t \in \mathbf{R} \text{ and } m \text{ is not identically zero}$$

$$\|m\|_{p_i} < \frac{\beta_{p_i}^{\mathrm{per}}}{b_i}$$

(5.57)

then for each continuous and 2π-periodic function $h : \mathbf{R} \to \mathbf{R}^n$, the boundary value problem

$$u''(t) + m(t)H_u(u(t)) = h(t), \ t \in \mathbf{R}, \ u(0) - u(2\pi) = u'(0) - u'(2\pi) = 0 \quad (5.58)$$

has a unique solution.

If in (5.57) we choose $p_i \neq \infty$, for some $1 \leq i \leq n$, then it is clear that the eigenvalues of the matrix $m(t)H_{uu}(u)$ in the boundary value problem (5.58) can cross an arbitrary number of eigenvalues of (5.43).

5.5 Stability of Linear Periodic Systems

As it was said in Chap. 1, the classical criterion (Lyapunov, Borg) on the stability of Hill's equation

$$u''(t) + p(t)u(t) = 0, \ t \in \mathbf{R} \quad (5.59)$$

with $p(\cdot)$ a T-periodic function asserts that if

$$p \in L_T(\mathbf{R}, \mathbf{R}) \setminus \{0\}, \ \int_0^T p(t) \, dt \geq 0, \ \int_0^T |p(t)| \, dt \leq \frac{4}{T} \quad (5.60)$$

then (5.59) is stable [19].

Condition (5.60) has been generalized in several ways. In particular, in [27], the authors provide optimal stability criteria by using L^r norms ($1 \leq r \leq \infty$) of the positive part p^+ of the function p.

Despite its undoubted interest, there are not many studies on the stability properties for systems of equations

$$u''(t) + P(t)u(t) = 0, \ t \in \mathbf{R}, \quad (5.61)$$

where the matrix function $P(\cdot)$ is T-periodic. A notable contribution was provided by Krein [13]. In this work, the author assumes that $P(\cdot) \in \Lambda_T$, where Λ_T is defined as

[Λ_T]

> The set of real $n \times n$ symmetric matrix valued function $P(\cdot)$, with continuous and T-periodic element functions $p_{ij}(t)$, $1 \leq i,j \leq n$, such that (5.61) has not nontrivial constant solutions and
>
> $$\int_0^T \langle P(t)k, k \rangle \, dt \geq 0, \ \forall \, k \in \mathbf{R}^n.$$

Krein proved that in this case, all solutions of the system (5.61) are stably bounded (see the precise definition of this concept below) if $\lambda_1 > 1$, where λ_1 is the smallest positive eigenvalue of the eigenvalue problem

$$u''(t) + \mu P(t)u(t) = 0, \ t \in \mathbf{R}, \ u(0) + u(T) = u'(0) + u'(T) = 0. \tag{5.62}$$

Conditions (5.60) and those given in [27] for scalar equations imply $\lambda_1 > 1$, but for systems of equations, and assuming $P(\cdot) \in \Lambda_T$, it is not easy to provide sufficient hypotheses which ensure the property $\lambda_1 > 1$(in [8, 9] the authors establish sufficient conditions for having $\lambda_1 > 1$ which involve L^1 restrictions on the spectral radius of some appropriate matrices, calculated by using the matrix $P(t)$. It is easy to check that, even in the scalar case, these conditions are independent from classical L^1 Lyapunov criterion (5.60)).

Next, we present several conditions which allow to prove that $\lambda_1 > 1$. These are given in terms of the L^p norm of appropriate functions $b_{ii}(t)$, $1 \leq i \leq n$, related to (5.61) through the inequality $P(t) \leq B(t)$, $\forall t \in [0, T]$, where $B(t)$ is a diagonal matrix with entries given by $b_{ii}(t)$, $1 \leq i \leq n$. These sufficient conditions are optimal in the sense explained in Remark 5.6 below.

From now on we assume that the matrix function $P(\cdot) \in \Lambda_T$. The system (5.61) is said to be stably bounded [13] if there exists $\varepsilon(P) \in \mathbf{R}^+$, such that all solutions of the system

$$u''(t) + Q(t)u(t) = 0, \ t \in \mathbf{R} \tag{5.63}$$

are bounded, for all matrix function $Q(\cdot) \in \Lambda_T$, satisfying

$$\max_{1 \leq i,j \leq n} \int_0^T |p_{ij}(t) - q_{ij}(t)| \, dt < \varepsilon.$$

The eigenvalue λ_1 (the smallest eigenvalue of (5.62)) has a variational characterization given by

$$\frac{1}{\lambda_1} = \max_{y \in G_T} \int_0^T \langle P(t)y(t), y(t) \rangle \, dt, \tag{5.64}$$

where

$$G_T = \left\{ y \in H^1(0, T) : y(0) + y(T) = 0, \ \sum_{i=1}^n \int_0^T (y_i'(t))^2 \, dt = 1 \right\}. \tag{5.65}$$

Theorem 5.6. *Let $P(\cdot) \in \Lambda_T$ be such that there exist a diagonal matrix $B(t)$ with continuous and T-periodic entries $b_{ii}(t)$, and $p_i \in [1, \infty]$, $1 \leq i \leq n$, satisfying*

$$P(t) \leq B(t), \ \forall t \in \mathbf{R}$$

$$\tag{5.66}$$

$$\|b_{ii}^+\|_{p_i} < \beta_{p_i}^{ant}, \ 1 \leq i \leq n$$

Then, the system (5.61) is stably bounded.

Proof. Let $y \in G_T$. Then by using Lemma 5.4, we have

$$\int_0^T \langle P(t)y(t), y(t) \rangle \, dt \leq \int_0^T \langle B(t)y(t), y(t) \rangle \, dt$$

$$\leq \sum_{i=1}^n \int_0^T b_{ii}(t)(y_i(t))^2(t) \, dt \leq \sum_{i=1}^n \|b_{ii}^+(t)\|_{p_i} \|y_i^2\|_{\frac{p_i}{p_i-1}}$$

$$\leq \sum_{i=1}^n \beta_{p_i}^{\text{ant}} \|y_i^2\|_{\frac{p_i}{p_i-1}} \leq \sum_{i=1}^n \int_0^T (y_i'(t))^2 \, dt = 1, \ \forall \ y \in G_T,$$

$$(5.67)$$

where

$$\frac{p_i}{p_i-1} = \infty, \ \text{if } p_i = 1$$

$$\frac{p_i}{p_i-1} = 1, \ \text{if } p_i = \infty.$$

We claim

$$\frac{1}{\lambda_1} < 1. \qquad (5.68)$$

In fact, (5.67) implies $\frac{1}{\lambda_1} \leq 1$. Now, if $\lambda_1 = 1$, let us choose $y(\cdot)$ as any nontrivial eigenfunction associated with $\mu = 1$ in (5.62), i.e.,

$$y''(t) + P(t)y(t) = 0, \ t \in \mathbf{R}, \ y(0) + y(T) = y'(0) + y'(T) = 0. \qquad (5.69)$$

Then some component, say y_j, is nontrivial and therefore $(\beta_{p_j}^{\text{ant}} - \|b_{jj}^+\|_{p_j})\|y_j^2\|_{\frac{p_j}{p_j-1}} > 0$. In addition, $(\beta_{p_i}^{\text{ant}} - \|b_{ii}^+\|_{p_i})\|y_i^2\|_{\frac{p_i}{p_i-1}} \geq 0, \ \forall \ i \neq j$, so that at least one of the inequalities in (5.67) is a strict inequality. This is a contradiction with (5.69).

Remark 5.6. Previous theorem is optimal in the following sense: for any given positive numbers γ_i, $1 \leq i \leq n$, such that at least one of them, say γ_j, satisfies

$$\gamma_j > \beta_{p_j}^{\text{ant}}, \ \text{for some } p_j \in [1, \infty] \qquad (5.70)$$

there exists a diagonal $n \times n$ matrix $P(\cdot) \in \Lambda_T$ with entries $p_{ii}(t)$, $1 \leq i \leq n$, satisfying $\|p_{ii}^+\|_{p_i} < \gamma_i$, $1 \leq i \leq n$ and such that the system (5.61) is not stable.

To see this, if γ_j satisfies (5.70), then there exists some continuous and T-periodic function $p(t)$, not identically zero, with $\int_0^T p(t) \, dt \geq 0$, and $\|p^+\|_{p_j} < \gamma_j$, such that the equation

$$w''(t) + p(t)w(t) = 0$$

is not stable (see Theorem 1 in [27]). If we choose

$$p_{jj}(t) = p(t), \ p_{ii}(t) = \delta \in \mathbf{R}^+, \ \text{if } i \neq j$$

with δ sufficiently small, then (5.61) is unstable.

Example 5.2.   Next we show a two-dimensional system where we provide sufficient conditions, checked directly by using the elements p_{ij} of the matrix $P(t)$, to assure that all hypotheses of the previous theorem are fulfilled. The example is based on a similar one for elliptic systems, shown by the authors in [6].

Let the matrix $P(t)$ be given by

$$P(t) = \begin{pmatrix} p_{11}(t) \ p_{12}(t) \\ p_{12}(t) \ p_{22}(t) \end{pmatrix}, \tag{5.71}$$

where

$$p_{ij} \in C_T(\mathbf{R}, \mathbf{R}), \ 1 \leq i, j \leq 2$$

[H1] $$p_{11}(t) \geq 0, \ p_{22}(t) \geq 0, \ \det P(t) \geq 0, \ \forall\, t \in \mathbf{R}$$

$$\det P(t) \neq 0, \ \text{for some } t \in \mathbf{R}$$

$C_T(\mathbf{R}, \mathbf{R})$ denotes the set of real, continuous, and T-periodic functions defined in $\mathbf{R}$. Then, if there exist $p_1, p_2 \in (1, \infty]$ such that

$$\|p_{11}\|_{p_1} < \beta_{p_1}^{\mathrm{ant}}, \ \left\| p_{22} + \frac{p_{12}^2}{\beta_{p_1}^{\mathrm{ant}} - \|p_{11}\|_{p_1}} \right\|_{p_2} < \beta_{p_2}^{\mathrm{ant}}. \tag{5.72}$$

Then (5.61) is stably bounded.

In fact, [**H1**] implies that the eigenvalues of the matrix $P(t)$ are both nonnegative, which implies that $P(t)$ is positive semi-definite. Also, since $\det P(t) \neq 0$, for some $t \in \mathbf{R}$, (5.61) has not nontrivial constant solutions. Therefore, $P(\cdot) \in \Lambda_T$.

Moreover, for a given diagonal matrix $B(t)$, with continuous entries $b_{ii}(t)$, $1 \leq i \leq 2$, the relation

$$P(t) \leq B(t), \ \forall\, t \in \mathbf{R} \tag{5.73}$$

is satisfied if and only if, $\forall\, t \in \mathbf{R}$ we have

$$b_{11}(t) \geq p_{11}(t), \ b_{22}(t) \geq p_{22}(t)$$

$$(b_{11}(t) - p_{11}(t))(b_{22}(t) - p_{22}(t)) \geq p_{12}^2(t). \tag{5.74}$$

In our case, we can choose

$$b_{11}(t) = p_{11}(t) + \gamma, \ b_{22}(t) = p_{22}(t) + \frac{p_{12}^2(t)}{\gamma}, \tag{5.75}$$

where γ is any constant such that

$$0 < \gamma < \beta_{p_1}^{\text{ant}} - \|p_{11}\|_{p_1}$$

$$\left(\frac{1}{\gamma} - \frac{1}{\beta_{p_1} - \|p_{11}\|_{p_1}}\right)\|p_{12}^2\|_{p_2} < \beta_{p_2}^{\text{ant}} - \|p_{22}\| + \frac{p_{12}^2}{\beta_{p_1}^{\text{ant}} - \|p_{11}\|_{p_1}}\|_{p_2}. \tag{5.76}$$

Then all conditions of Theorem 5.6 are fulfilled and consequently (5.61) is stably bounded.

5.6 Partial Differential Equations

This section will be concerned with vector boundary value problems of the form

$$\Delta u(x) + A(x)u(x) = 0, \; x \in \Omega, \quad \frac{\partial u(x)}{\partial n} = 0, \; x \in \partial\Omega. \tag{5.77}$$

Here $\Omega \subset \mathbf{R}^N$, $N \geq 2$ is a bounded and regular domain, $\dfrac{\partial}{\partial n}$ is the outer normal derivative on $\partial\Omega$, and $A \in \Lambda_*$, where Λ_* is defined as

[Λ_*]
> The set of real $n \times n$ symmetric matrix valued function $A(\cdot)$, with continuous element functions $a_{ij}(x)$, $1 \leq i, j \leq n$, $x \in \overline{\Omega}$, such that (5.77) has not nontrivial constant solutions and
>
> $$\int_\Omega < A(x)k, k > dx \geq 0, \; \forall \, k \in \mathbf{R}^n. \tag{5.78}$$

We will consider solutions $u \in (H^1(\Omega))^n$.

Next result may be proved by using the same ideas as in Theorem 5.2.

Theorem 5.7. *Let $A(\cdot) \in \Lambda_*$ be such that there exist a diagonal matrix $B(x)$ with continuous entries $b_{ii}(x)$, and values $p_i \in (N/2, \infty]$, $1 \leq i \leq n$, which fulfill*

$$A(x) \leq B(x), \; \forall \, x \in \overline{\Omega}$$

$$\|b_{ii}^+\|_{p_i} < \beta_{p_i}, \; 1 \leq i \leq n, \tag{5.79}$$

where β_{p_i} is defined in 4.5.

Then, there exists no nontrivial solution of the vector boundary value problem (5.77).

Remark 5.7. As in the ordinary case, the previous theorem is optimal in the sense of Remark 5.1. Moreover, by using the previous theorem, it is possible to obtain a corollary similar to Corollary 5.1 and a result similar to the exposed in Remark 5.3, which involves the spectral radius $\rho(x)$ of the matrix $A(x)$ and the norm $\|\rho^+\|_p$. The unique difference with the ordinary case is that, for elliptic systems, $p \in (N/2, \infty]$.

Finally, we exhibit a result on the existence and uniqueness of solutions of nonlinear resonant problems, which proof can be obtained by using the same ideas as in Theorem 5.5 (see [6]) for the details.

Theorem 5.8. *Let $\Omega \subset \mathbf{R}^N$ ($N \geq 2$) be a bounded and regular domain and $G : \overline{\Omega} \times \mathbf{R}^n \to \mathbf{R}$, $(x, u) \to G(x, u)$ satisfying:*

1. *a. The mapping $u \to G(x, u)$ is of class $C^2(\mathbf{R}^n, \mathbf{R})$ for every $x \in \overline{\Omega}$.*
   b. The mapping $x \to G(x, u)$ is continuous on $\overline{\Omega}$ for every $u \in \mathbf{R}^n$.
2. *There exist continuous matrix functions $A(\cdot)$, $B(\cdot)$, with $B(x)$ diagonal and with entries $b_{ii}(x)$, and $p_i \in (N/2, \infty]$ $1 \leq i \leq n$, such that*

$$\left. \begin{array}{c} A(x) \leq G_{uu}(x, u) \leq B(x) \text{ in } \overline{\Omega} \times \mathbf{R}^n \\[2mm] \|b_{ii}^+\|_{p_i} < \beta_{p_i}, \ 1 \leq i \leq n \\[2mm] \int_{\Omega} <A(x)k, k> dx > 0, \ \forall \, k \in \mathbf{R}^n \setminus \{0\} \end{array} \right\}. \tag{5.80}$$

Then system

$$\left. \begin{array}{c} \Delta u(x) + G_u(x, u(x)) = 0, \ x \in \Omega \\[2mm] \frac{\partial u(x)}{\partial n} = 0, \ \ x \in \partial\Omega \end{array} \right\} \tag{5.81}$$

has a unique solution.

References

1. Ahmad, S.: An existence theorem for periodically perturbed conservative systems. Mich. Math. J. **20**, 385–392 (1973)
2. Bates, P.W.: Solutions of nonlinear elliptic systems with meshed spectra. Nonlinear Anal. **4**, 1023–1030 (1979)
3. Brown, K.J., Lin, S.S.: Periodically perturbed conservative systems and a global inverse function theorem. Nonlinear Anal. **4**, 193–201 (1980)
4. Cañada, A., Montero, J.A., Villegas, S.: Liapunov-type inequalities and Neumann boundary value problems at resonance. Math. Inequal. Appl. **8** 459–475 (2005)
5. Cañada, A., Villegas, S.: Stability, resonance and Lyapunov inequalities for periodic conservative systems. Nonlinear Anal. **74**, 1913–1925 (2011)
6. Cañada, A., Villegas, S.: Matrix Lyapunov inequalities for ordinary and elliptic partial differential equations. Topological Methods Nonlinear Anal. **45**, 326–329 (2015)
7. Choquet, G.: Topology. Academic, New York (1966)

8. Clark, S., Hinton, D.: A Liapunov inequality for linear hamiltonian systems. Math. Inequal. Appl. **1**, 201–209 (1998)
9. Clark, S., Hinton, D.: Positive eigenvalues of second order boundary value problems and a theorem of M.G. Krein. Proc. Am. Math. Soc. **130**, 3005–3015 (2002)
10. Deimling, K.: Nonlinear Functional Analysis. Springer, Berlin (1985)
11. Fonda, A., Mawhin, J.: Iterative and variational methods for the solvability of some semilinear equations in Hilbert spaces. J. Differ. Equ. **98**, 355–375 (1992)
12. Hartman, P.: Ordinary Differential Equations. Wiley, New York (1964)
13. Krein, M.G.: On tests for stable boundedness of solutions of periodic canonical systems. Am. Math. Soc. Translat. Ser. (2) **120**, 71–110 (1983)
14. Lazer, A.C.: Application of a lemma on bilinear forms to a problem in nonlinear oscillations. Proc. Am. Math. Soc. **33**, 89–94 (1972)
15. Lang, S.: Algebra. Addison-Wesley, Reading, MA (1965)
16. Lang, S.: Analysis.II. Addison-Wesley, Reading, MA (1969)
17. Lazer, A.C., Sánchez, D.A.: On periodically perturbed conservative systems. Mich. Math. J. **16**, 193–200 (1969)
18. Lee, E.B., Markus, L.: Foundations of Optimal Control Theory. Wiley, New York (1967)
19. Magnus, W., Winkler, S.: Hill's Equation. Dover Publications, New York (1979)
20. Mawhin, J.: Contractive mappings and periodically perturbed conservative systems. Arch. Math. (Brno) **12**, 67–74 (1976)
21. Mawhin, J.: Conservative systems of semilinear wave equations with periodic-Dirichlet boundary conditions. J. Differ. Equ. **42**, 116–128 (1981)
22. Pinasco, J.P.: Lyapunov-Type Inequalities. With Applications to Eigenvalue Problems. Springer Briefs in Mathematics. Springer, New York (2013)
23. Reid, W.T.: A matrix Liapunov inequality. J. Math. Anal. Appl. **32**, 424–434 (1970)
24. Reid, W.T.: A generalized Liapunov inequality. J. Differ. Equ. **13**, 182–196 (1973)
25. Reid, W.T.: Interrelations between a trace formula and Liapunov type inequalities. J. Differ. Equ. **23**, 448–458 (1977)
26. Ward, J.R.: The existence of periodic solutions for nonlinear perturbed conservative systems. Nonlinear Anal. **3**, 697–705 (1979)
27. Zhang, M., Li, W.: A Lyapunov-type stability criterion using L^α norms. Proc. Am. Math. Soc. **130**, 3325–3333 (2002)
28. Zhang, M.: Certain classes of potentials for p-Laplacian to be non-degenerate. Math. Nachr. **278**, 1823–1836 (2005)

Index

© The Author(s) 2015
A. Cañada, S. Villegas, *A Variational Approach to Lyapunov Type Inequalities*,
SpringerBriefs in Mathematics, DOI 10.1007/978-3-319-25289-6